SUSTAINABLE ENERGY POLICIES FOR EUROPE –
TOWARDS 100% RENEWABLE ENERGY

Sustainable Energy Developments

Series Editor

Jochen Bundschuh
University of Southern Queensland (USQ), Toowoomba, Australia
Royal Institute of Technology (KTH), Stockholm, Sweden

ISSN: 2164-0645

Volume 6

Sustainable Energy Policies for Europe

Towards 100% Renewable Energy

Rainer Hinrichs-Rahlwes

European Renewable Energy Council and *European Renewable Energies Federation, Brussels, Belgium,* and *German Renewable Energy Federation, Berlin, Germany*

with contributions from:

Christine Lins
REN21, Paris, France

Jan Geiss
EUFORES, Brussels, Belgium

Markus Kahles & Thorsten Müller
SUER, Würzburg, Germany

CRC Press
Taylor & Francis Group
Boca Raton   London   New York   Leiden

CRC Press is an imprint of the
Taylor & Francis Group, an **informa** business

A BALKEMA BOOK

Cover photo

Martin Schulz – EP President presiding the plenary sitting in Strasbourg, general view of the plenary chamber in Strasbourg, January 17, 2013

© European Union 2013 – Source: EP

CRC Press/Balkema is an imprint of the Taylor & Francis Group, an informa business

© 2013 Taylor & Francis Group, London, UK

Typeset by MPS Limited, Chennai, India
Printed and bound in The Netherlands by PrintSupport4U, Meppel

Library of Congress Cataloging-in-Publication Data

Sustainable energy policies for Europe : towards 100% renewable energy /
   Rainer Hinrichs-Rahlwes; with contributions from Christine Lins, Jan Geiss,
   Markus Kahles & Thorsten Müller.
      pages cm. — (Sustainable energy developments, ISSN 2164-0645; volume 6)
   Includes bibliographical references and index.
   ISBN 978-0-415-62099-4 (hardback : alkaline paper) — ISBN 978-0-203-12022-4 (e-book)
   1. Energy policy—Europe.   2. Renewable energy sources—Government policy—Europe.
I. Hinrichs-Rahlwes, Rainer.
   HD9502.E82S87   2013
   333.79'4094—dc23

                                                                                    2013021552

Published by: CRC Press/Balkema
                P.O. Box 11320, 2301 EH, Leiden, The Netherlands
                e-mail: Pub.NL@taylorandfrancis.com
                www.crcpress.com – www.taylorandfrancis.com

ISBN: 978 0 415 62099 4 (Hardback)
ISBN: 978 0 203 12022 4 (e-book)

About the book series

Renewable energy sources and sustainable policies, including the promotion of energy efficiency and energy conservation, offer substantial long-term benefits to industrialized, developing and transitional countries. They provide access to clean and domestically available energy and lead to a decreased dependence on fossil fuel imports, and a reduction in greenhouse gas emissions.

Replacing fossil fuels with renewable resources affords a solution to the increased scarcity and price of fossil fuels. Additionally it helps to reduce anthropogenic emission of greenhouse gases and their impacts on climate change. In the energy sector, fossil fuels can be replaced by renewable energy sources. In the chemistry sector, petroleum chemistry can be replaced by sustainable or green chemistry. In agriculture, sustainable methods can be used that enable soils to act as carbon dioxide sinks. In the construction sector, sustainable building practice and green construction can be used, replacing for example steel-enforced concrete by textile-reinforced concrete. Research and development and capital investments in all these sectors will not only contribute to climate protection but will also stimulate economic growth and create millions of new jobs.

This book series will serve as a multi-disciplinary resource. It links the use of renewable energy and renewable raw materials, such as sustainably grown plants, with the needs of human society. The series addresses the rapidly growing worldwide interest in sustainable solutions. These solutions foster development and economic growth while providing a secure supply of energy. They make society less dependent on petroleum by substituting alternative compounds for fossil-fuel-based goods. All these contribute to minimize our impacts on climate change. The series covers all fields of renewable energy sources and materials. It addresses possible applications not only from a technical point of view, but also from economic, financial, social and political viewpoints. Legislative and regulatory aspects, key issues for implementing sustainable measures, are of particular interest.

This book series aims to become a state-of-the-art resource for a broad group of readers including a diversity of stakeholders and professionals. Readers will include members of governmental and non-governmental organizations, international funding agencies, universities, public energy institutions, the renewable industry sector, the green chemistry sector, organic farmers and farming industry, public health and other relevant institutions, and the broader public. It is designed to increase awareness and understanding of renewable energy sources and the use of sustainable materials. It aims also to accelerate their development and deployment worldwide, bringing their use into the mainstream over the next few decades while systematically replacing fossil and nuclear fuels.

The objective of this book series is to focus on practical solutions in the implementation of sustainable energy and climate protection projects. Not moving forward with these efforts could have serious social and economic impacts. This book series will help to consolidate international findings on sustainable solutions. It includes books authored and edited by world-renowned scientists and engineers and by leading authorities in in economics and politics. It will provide a valuable reference work to help surmount our existing global challenges.

Jochen Bundschuh
(Series Editor)

Editorial board

Table of contents

Foreword by Günther Oettinger

Back in 2007, the European Union and its Member States set themselves ambitious goals: the reduction of greenhouse gas emissions by 20%, increase energy efficiency by 20%, and finally a share of renewable energy sources by 20% in the overall EU energy mix. These goals also feature as headline targets of the European 2020 strategy for growth, since they contribute to Europe's industrial innovation and technological leadership as well as reducing greenhouse gas emissions.

Renewable energy sources play a very important role in tackling the security of supply, competitiveness and sustainability challenges. They reduce the dependency on fossil fuel imports and they offer opportunities for new market players.

By now, renewable energies have become "a major player in the European energy market" as acknowledged in the Commission Communication "Renewable Energy: a major player in the European energy market" and renewables currently account for 13% of final energy consumption.

But with size comes responsibility–a necessity that is further accelerated by the current difficult economic climate. Dedicated support schemes for renewable energies have been instrumental in bringing about technological innovation and economies of scale that led to major cost reductions, and strong growth. But support schemes also have to be flexible enough to adapt to price decreases as technologies mature and avoid overcompensation. Only support schemes that are well designed, i.e. flexible, degressive, and market based, can integrate higher renewable shares in a cost-efficient way to reach our set targets. The Commission has been calling on Member States to follow these principles. Additional efforts are also needed to integrate renewable energy into the market. Member States should also enhance their cooperation when developing their renewable energy sources in order to make use of the most cost-effective options. Now is the time to translate political commitments of making the internal market work into daily reality for all power sources. The existing infrastructure and grid rules need to be modernised. The European infrastructure package as well as our on-going work on harmonisation of network codes are important steps in that direction. These measures will help to accommodate in a European context large shares of renewable energy. But the power system of the future will also have to include more active consumer participation in the market through demand response and decentralised generation, flexible and smart grids, storage, as well as a deeper integration of liquid markets in all timeframes.

Higher shares of renewable energy are only sustainable and affordable under these conditions. Adequate support schemes will continue to play an important role, but market mechanisms will increasingly have to drive investments in renewable energy sources.

In all of this, investor certainty is key. Investments in the energy sector require long lead times and thus we are looking at defining our climate and energy policy targets until 2030. The new challenges and the lessons learned from the past will shape this decision. Whatever the outcome of our reflections, there is no doubt that the further development of renewable energy will continue to play a key role in the future energy policy framework.

Günther H. Oettinger
Commissioner for Energy, European Commission

Foreword by Fiona Hall

The EU's climate and energy package with the 2020 targets for greenhouse gas emissions reduction, energy efficiency and renewable energy share in gross final energy consumption is an important milestone on the road towards a fully sustainable energy system in Europe. The political agreement of the March 2007 Council Meeting was transformed into concrete legislation and eventually materialized in the form of the Renewable Energy Directive, the Effort Sharing Decision, the ETS Directive and the CCS Directive. This did not happen without major efforts of Members of the European Parliament in close cooperation with stakeholders, Member States' governments and the European Commission. Only due to perseverance and decisive action of those MEPs who had been striving for more ambitious greenhouse gas reductions, increased energy efficiency and an improved renewable energy policy framework for years was it possible to develop and promote a consistent framework for the implementation of the 2020 targets.

The climate and energy package as it stands today was only possible because Members of the European Parliament together with NGOs and stakeholders managed to convincingly promote and defend the coherence of the Council agreement on binding EU and national targets and in particular the integrity of national support systems for renewable energy. It is thanks to the Rapporteur and the Shadow Rapporteurs of the European Parliament that the systematic approach of the Renewable Energy Directive was developed and fine-tuned: Members States are responsible for reaching their binding 2020 targets and therefore their control of the national support systems must not be affected. Member States may, however, on a voluntary basis use cooperation mechanisms for cost-effective target implementation. The key role of the Member States in designing and improving support policies is balanced by the right and the obligation of the European Commission to thoroughly check the National Renewable Energy Action Plans (NREAPs) for coherence and for compliance with the 2020 target achievement.

Other parts of the climate and energy package are – unfortunately – less strong and less effective than the Renewable Energy Directive. The Emissions Trading Directive introduced Europe-wide auctioning of emissions allowances, which should have been a major step towards effective carbon pricing and uptake of low carbon technology. However, the ETS so far was not able to cope with the large amounts of excess allowances of the first trading periods and the subsequent economic downturn. As of today, only short-term "backloading" of some of the excess allowances is on the table but still subject to negotiation between the Parliament and the Council. Fundamental improvement of the ETS has not yet been attempted.

Efficient policies and targets for energy efficiency still have a long way to go. The revised Energy Efficiency Directive is an important step in this direction. However, a breakthrough in energy efficiency policies will require a commitment by Member States beyond 2020, and stronger instruments for implementation and financing – another big task for the years to come.

Eventually, the key role of renewable energy development and deployment needs to be further strengthened – for successful greenhouse gas emissions reduction, and even more so for energy supply security, economic growth and job creation. The EU therefore needs to agree on a new climate and energy package with a view to setting ambitious and binding 2030 targets. I very much welcome the European Commission's Green Paper on the 2030 framework for climate and energy policies. All the major questions are addressed in the paper, but now we need to develop, enact and eventually implement an integrated 2030 framework with binding targets for carbon reduction, energy efficiency and renewables.

Given that the Commission has provided the framework and given that many Member States are still hesitating to commit to a new set of ambitious and binding targets for 2030, I see once again a strong and important role for the European Parliament in facilitating the process which should, sooner rather than later, end in a political agreement followed by comprehensive legislation. Agreement on a 2030 framework with ambitious and binding targets will not only help implementing the 2020 targets more smoothly and more cost-efficiently, it will also re-establish investors' confidence for investment in sustainable energy until and beyond 2020. Paving the way for fast deployment of renewables and increased energy efficiency is a major driver for a fully sustainable energy future for Europe. I am sure that the European Parliament will play a major part in facilitating an ambitious 2030 framework.

Fiona Hall
Member of the European Parliament (MEP)
Leader of UK Liberal Democrats
Vice-President EUFORES

Author's preface

The European Commission's Green Paper *"A 2030 framework for climate and energy policies"* (European Commission 2013a) and the related public consultation officially started the process. In autumn 2013, the Commission plans to present a proposal for a post-2020 policy framework and targets for the next decade. The European Parliament has decided to discuss and vote an initiative report in order to develop a detailed position on the 2030-framework before the end of 2013. The Heads of State and Government have envisaged for the European Council in March 2014 to agree on the level of ambition in climate and energy policies until 2030 and on the next steps to taken including the outline of a policy framework for the next decade. Hence, another crucial phase for the EU's sustainable energy future has begun.

So far, the European Union and most of the Member States have successfully developed and implemented supportive frameworks, in particular for renewable energy development and deployment. From the first support programmes on EU level in the mid-1980s, followed by the White Paper 1997, the Campaign for Take-off 1999, the Renewables Electricity Directive 2001, the Biofuels Directive 2003, and eventually the Climate and Energy Package 2009, the policy framework was smoothly developed and fine-tuned. The transition towards a fully sustainable energy system is well prepared and facilitated by supportive legal, regulatory and administrative frameworks. However, in recent months, some Member states have changed their policies. They have reduced their ambition level and changed policy frameworks. Some of them have even implemented retrospective changes, which affect existing installations.

In the Energy Roadmap 2050, the European Commission underlined that decarbonizing the energy sector until 2050 is both necessary to meet the greenhouse gas reduction targets and feasible at reasonable costs – with renewable energy providing the major share of Europe's energy supply in all decarbonization scenarios and delivering benefits for economy and environment. Others have provided evidence that by 2050 a complete transition towards a renewables based energy system is necessary and possible.

Scenarios, however, are projections, which do not come true without enabling policies. Although it is true that renewable energy will certainly continue growing due to further decreasing costs and benefits provided. Nevertheless, Europe's energy supply is still heavily depending on policy decision. Whether investment decisions will be deviated to risky and costly fossil and nuclear energy sources or whether most of the investment will be made for renewable energy, energy efficiency and related infrastructure will be significantly influenced by policy decisions to be taken in the near future. This is why a timely agreement on a stable, reliable and ambitious climate and energy framework with a set of equally ambitious and binding targets for renewable energy, energy efficiency and greenhouse gas reduction – on EU-level and on Member State level – is of crucial importance.

This book describes the development of Europe's climate and energy policies from the early days until mid-2013, when the manuscript was finalized. It provides background information about the development of the EU's policies in the area until today. By analysing scenarios up to 2050 and recent policy discussions about the future of Europe's energy supply, it provides a solid background for understanding the ongoing policy debate and the upcoming controversies. The book argues for a fully sustainable energy system, for facilitating a smooth transition towards 100% renewable energy. Swift implementation of enabling policies will reduce costs and provide plenty of benefits – economic growth, supply security and future oriented jobs. To achieve the objective of a fully renewables based energy system ambitious mid-term targets for renewables and energy efficiency and stable and predictable framework conditions need to be in place. The leaders of the European Union should seize the opportunity and agree on an ambitious and realistic climate and energy package for 2030. This should be one of the outstanding priorities for 2014 and early 2015.

Rainer Hinrichs-Rahlwes, 2 July 2013

About the author

Rainer Hinrichs-Rahlwes (*1954, Germany) is currently the President of the European Renewable Energy Council (EREC), the Brussels based umbrella organisation of the European renewable energy sector. He is the President of EREC's member association, the European Renewable Energies Federation (EREF), the voice of independent producers of energy from renewable sources, and he is a Board Member and the spokesperson for European and International Affairs of the German Renewable Energy Federation (BEE), the national umbrella orginsation of the renewable energy sector.

He is closely engaged in European policy development for renewable energies in the European Union as well as in his home country, keeping close contacts with government representatives, parliamentarians, European Commission and other stakeholders. He is convinced that a complete shift of our energy system to renewable energy is necessary for the sake of energy security and climate protection and that it is technically and economically feasible – much faster and less costly than supporters and beneficiaries of conventional and nuclear energy are trying to make believe. Rainer Hinrichs-Rahlwes has delivered speeches and presentations and participated in panel discussions all over the world – on behalf of the organisations he is representing or advising, and as an independent consultant providing policy advice and knowledge about sustainable renewable energy development and policies for scaling up renewables on local, national, regional and global level in order to facilitate their becoming the mainstream energy sources already in the near future.

Representing EREC he is a member of the Renewable Energy Industry Advisory Board (RIAB) of the International Energy Agency (IEA) and a member of the Steering Committee of the global Renewable Energy Policy Network for the 21st Century (REN21) with headquarters in Paris (France), which was founded as an outcome of the first *"International Renewable Energy Conference" (IREC)*, the renewables2004-conference in Bonn. He is also a member of the WREN-Council, the advisory structure of the World Renewable Energy Network/Congress.

Before engaging with the renewable energy sector in Germany and in Europe, from November 1998 to December 2005, Rainer Hinrichs-Rahlwes was a Director General in the German Federal Ministry for the Environment, Nature Conservation and Nuclear Safety (BMU), in charge of renewable energies, climate protection and various other dossiers. As a representative of BMU, he was one of the two chairmen of the International Steering Committee preparing the first IREC, the renewables2004 in Bonn. After the conference, until he left the ministry at the end of 2005, he served as BMU's representative and a founding co-chair and later a member of the Bureau of the Global Policy Network, now known as REN21.

About the contributors

Christine Lins was appointed as Executive Secretary of REN21, the Renewable Energy Policy Network of the 21st Century, in July 2011. REN21 is a global public-private multi-stakeholder network on renewable energy regrouping international organizations, governments, industry associations, science and academia as well as NGOs working in the field of renewable energy. REN21 has its headquarters at UNEP, the United Nations Environment Programme in Paris/France. Between 2001 and 2011, she served as Secretary General of the European Renewable Energy Council, the united voice of Europe's renewable energy industry. She has more than 17 years of working experience in the field of renewable energy sources. Previously, she worked in a regional energy agency in Austria promoting energy efficiency and renewable energy sources. Christine Lins holds a Masters degree in international economics and applied languages.

Dr. Jan Geiss is Secretary General of EUFORES – The European Forum for Renewable Energy Sources, a network of Members of the European Parliament and the EU28 national parliaments promoting renewable energy and energy efficiency policy and markets in the European Union. He has been running the Brussels based organisation since 2006. Before becoming the Secretary General, he was Policy Advisor and Managing Director of EUFORES. Since 2011, he is also the President of the Renewable Energy House in Brussels. Since 2012, he is a member of the Business Council of the German Foundation for the Environment. 1999–2012, he was Chair of the Board of the Sustainable Development Forum Germany. He holds a PhD in Political Science and Economics from the University of Passau, Germany in the field of "Renewable Energy and Energy Efficiency Service Contracting". He finished his studies of International Cultural and Business Management in 1997.

Thorsten Müller is research director and chairman of the executive board of the Foundation on Environmental Energy Law (*Stiftung Umweltenergierecht* – SUER) in Würzburg (Germany). After studying law from 1996 to 2001 in Würzburg and Salzburg (Austria) and a legal traineeship at the OLG Celle (higher regional court, Germany) he worked as counsel for the German Ministry for the Environment, Nature Conservation and Nuclear Safety on 2003 and 2004-amendment of the Renewable Energies Act (EEG). Starting November 2004 he also worked as research assistant at the chair of Prof. Dr. Schulze-Fielitz for Public, Environmental and Administrative Law at the University of Würzburg. From 2006 to 2011 he was the head of the *Forschungsstelle Umweltenergierecht* (Research Centre for Energy and Environmental Law) in cooperation with the University of Würzburg. He published in legal journals and contributed to law commentaries and handbooks. He co-edits *"Zeitschrift für Neues Energierecht – ZNER"* (New Energy Law Journal), *"Renewable Energy Law and Policy Review (RELP)"* and *"Schriften zum Umweltenergierecht"* (Writings on Renewable Energy Law) published by Nomos-Verlag. His focus in legal research lies on European and national renewable energies law as well as energy efficiency law and the interactions between the different legal instruments. Repeatedly he acted as legal expert in hearings held by the German Bundestag as well as regional parliaments (Landtage) and governments.

Since 2010 **Markus Kahles** has been research assistant at the Foundation for Environmental Energy Law. His research focus lies on European renewable energies law in general and especially on state aid law and the law of free movement of goods. From 2004 to 2010 he studied law combined with accompanying studies in European law at the University of Würzburg (Germany) and the University of Bergen (Norway). At the moment he is doing his legal traineeship at the OLG Bamberg (higher regional court) and writing his doctoral thesis in the field of European renewable energy law.

Acknowledgements

I would like to thank all those who encouraged me to carry on with this book, although it sometimes seemed – with all the workload I was facing in the policy arena in Brussels and elsewhere – that I could not find the time for doing so. I wish to extend my special thanks to the contributing authors, without whom I could not have finished the project and who provided their insights and cooperation. I would also like to thank Jochen Bundschuh, the series editor, who convinced me to write the book and who helped checking and editing the final text.

Eventually, I am very grateful to my family, particularly to my wife, who had to bear part of the burden of my writing the book, because working on the book reduced the time I could spend with her.

Rainer Hinrichs-Rahlwes
May 2013

Conventions

We use (in uppercase) for:

Parliament: European Parliament
Union: European Union
Council: European Council
Commission: European Commission
Member States: for EU Member States
State Aid: as a technical term in EU legislation

Abbreviations:

PV: Photovoltaics
RES: Renewable Energy Sources
MEP: Member of European Parliament
ETS: Emissions Trading System
CCS: Carbon Capture and Storage

The importance of sustainable energy policies for Europe – an introduction

Rainer Hinrichs-Rahlwes

1 CLIMATE CHANGE: THE CHALLENGE TO BE MET

Climate change is anthropogenic, man made. Since the beginning of industrialisation, greenhouse gas emissions and resulting atmospheric concentration have increased dramatically. Longstanding concentrations below 300 ppm (parts per million) increased rapidly, since industrialisation began. Business-as-usual scenarios see the world quickly approaching the threshold of 450 ppm CO_2 equivalent, which scientists consider to be the upper limit of greenhouse gas saturation of the atmosphere – if we want to limit global warming to a maximum of 2°C (or 1.5°C) compared to pre-industrial levels and thus prevent most catastrophic impacts. In recent years, growing evidence was provided that 450 ppm might still be much too high for achieving the 2°-target (and even more so for targeting a lower temperature rise, which more vulnerable countries hold to be necessary). A new target of limiting greenhouse gas concentration to a maximum of 350 ppm (or lower) is being introduced in the scientific discussion and in the political process. Whereas the lower limit seems to become scientific common sense, it is far from being political consensus and from becoming the new target for the international multi-lateral climate process[1].

Whereas evidence is growing that greenhouse gas reduction is even more urgent than it was considered a few decades or only a few years ago, global greenhouse gas emissions are further increasing – particularly in the developing world. Efficiency gains in the industrialised world are being more than overcompensated by increasing energy needs for new technologies. Meanwhile China has overtaken the United States of America as the world's biggest greenhouse gas emitting country. Other major developing economies – e.g. India, Brazil, Mexico, and the ASEAN countries – are rapidly increasing their greenhouse gas emissions. On a per capita basis, they still have much lower emissions than the industrialised countries in Europe and in particular in North America. But the trend is clear. The more countries reach a certain level of development and resulting industrialisation, the more greenhouse gas emissions will increase – in absolute terms and potentially also on a per capita basis.

In line with the United Nations' goals for poverty alleviation there are good reasons for further increasing efforts to facilitate access to sustainable energy for all. The need to provide access to energy for the developing world is another strong reason for the industrialised countries to reduce their own greenhouse gas emissions. In order to achieve the United Nations' development goals and at the same time limit atmospheric greenhouse gas concentrations to below 450 ppm (or 350 ppm or less), scientific debates and political declarations seem to agree that the industrialised countries have to reduce their emissions by 80 to 95% until 2050 and that effective measures to

---

[1]This book will not evaluate these facts and discussions in detail. It is not about assessing global climate change from a scientific point of view, nor the global climate process a major focus here. There are other authors and experts, e.g. in the framework of the IPCC, who have analysed and assessed the process and the outcomes. But climate change is a major driver for sustainable energy policies globally. Therefore, I believe it is useful to briefly elaborate on the global context in which sustainable energy policies for Europe are developing.

achieve this target have to be in place in the next few years. Otherwise, due to lock-in effects from existing installations and those under construction, the chances for effectively limiting global warming to a tolerable (?) level would drastically decrease – or vanish completely.

2   CLIMATE NEGOTIATIONS: STAGNATION FOR YEARS

Whereas the global problem is evident, global solutions are not. The Intergovernmental Panel on Climate Change (IPCC)[2] has a high reputation, scientifically and as an advisory body to the international climate process. The global arena for greenhouse gas mitigation is there; international climate negotions have been conducted for years, with ad-hoc consultations, with subsidiary bodies and with annual Conferences (COP/MOP). Discussions and negotions are ongoing. The structures of the United Nations Framework Convention on Climate Change (UNFCCC)[3] are established and many people are working hard to achieve some progress. The Kyoto Protocol[4] is still in place; most of the Member States seem to be willing to negotiate an extension with new targets and improved structures. However, after the 2009 COP15/MOP5 in Copenhagen[5], which environmental NGOs world-wide as well as many governments had overstated to become the breakthrough for a new and ambitious global compact on climate change, sobriety has replaced high expectations towards the multi-lateral process.

After three more conferences, 2010 in Cancún/Mexico[6], 2011 in Durban/South Africa[7] and 2012 in Doha/Quatar[8], it is evident that basic questions are still open. Financial needs are encoded, but actually providing financing is more unclear than before. The Doha agreements[9] basically mean that there will be conferences in 2013 and 2014, and they highlight the participants' determination *"to adopt a protocol, another legal instrument or an agreed outcome with legal force under the Convention applicable to all Parties"* in 2015, which shall become effective and be implemented from 2020 onwards. Bearing in mind how much effort was needed for the participants of the Doha conference to agree on this minimum consensus, it is difficult to imagine how an agreement can be achieved in 2015 which would significantly contribute to alleviating climate change.

Some civil society representatives and politicians[10] have drawn their own conclusions from these many years of negations. Some suggested boycotting future climate talks, because they consider them useless and even delaying or preventing solutions. It may be questioned whether time and money is reasonably and efficiently spent in the international climate process, trying to find minimum consensus. But the climate process has had some impact in raising awareness about the urgency of mitigating global warming. And it will continue, at least until 2015. There will be an agreement at COP21 in 2015 – despite growing doubts about the willingness and ability of major global players to reach consensus in time and ambitious enough to limit global warming to 2 or even 1.5 degrees. The new agreement will ratify the minimum consensus reached by then – and it will be far from sufficient to stop global warming. There is, however, still reason

---

[2] http://www.ipcc.ch/ for details about IPCC's mission, structure and activities.
[3] http://unfccc.int/ for information about the convention's work, relevant documents, meetings.
[4] http://unfccc.int/kyoto_protocol/items/2830.php
[5] http://unfccc.int/meetings/copenhagen_dec_2009/meeting/6295.php
[6] http://unfccc.int/meetings/cancun_nov_2010/items/6005.php
[7] http://unfccc.int/meetings/durban_nov_2011/meeting/6245.php
[8] http://unfccc.int/meetings/doha_nov_2012/meeting/6815.php
[9] http://unfccc.int/2860.php#decisions
[10] One of most outspoken criticics of the international climate process, German Greens-MP, Hans-Josef Fell, suggested to boycott future climate negations, because he considers the conferences as a major barrier against successful and effective climate protection measures: http://www.hans-josef-fell.de/content/index.php?option=com_content&view=article&id=573:doha-minimalkonsens-fuer-schnelle-aufheizung-der-erde&catid=24:schlagzeilen&Itemid=73

to believe that it will at least mitigate some of the worst consequences of climate change for the poorest and the most vulnerable, but action cannot wait until then. And waiting is not necessary. It is technically possible to replace greenhouse gas emitting resources and technologies by clean and sustainable renewable energy – within a few years. And it is economically reasonable to do so, because costs for renewables are constantly decreasing, whereas conventional sources are depleting – with resulting volatile and increasing costs. The earlier clear and unambiguous policy decisions are taken, the more effective, successful and economically beneficial the energy turnaround will be. Every country and every region can start. And many already did.

3   RENEWABLE ENERGY: THE SOLUTION AT HAND

The energy sector is a major source of greenhouse gas emissions. Rapidly growing energy needs for industrial processes and for the transport sector, for domestic heating and cooling and for communication are main drivers for accelerated power plant deployment all over the world. More and more countries have started to replace old fossil power plants by renewable energy, such as hydro, wind, solar, bioenergy and geothermal. This includes some of the big developing countries – namely China and India, but also Brazil, Mexico and others – which are implementing ambitious programmes for renewable energy to provide growing shares of the countries energy needs. But they continue building new conventional power plants, which add to existing emitters and historical greenhouse gas emissions. Recent studies[11] suggest that there is a significant risk of locking-in greenhouse gas emissions from existing and newly planned fossil power plants, to an extent that could soon exceed the 450 ppm threshold.

Given the actual standstill of international climate negotiations and the unlikeliness of a sufficiently ambitious consensus for a post-2020 global climate agreement unilateral action on local, regional, national and continental level is increasingly relevant – and will prove to be beneficial for those who undertake it. Local activities for greenhouse gas reduction often include ambitious targets for high shares of renewable energy in the near future – on a calculatory basis via community investment in large scale projects and/or in the form of cooperative or community action for using renewable energy. This not only creates and secures hundreds of thousands of jobs and a clean environment. In 2011, more than five million jobs worldwide were related to renewable energy. For details see Chapter 10 and the REN21's Renewables Global Status Report (GSR 2012)[12]. Development and deployment of renewables triggers billions of investment in clean energy and thus generates tax income for public budgets. In 2011, global investment amounted to 257 billion US-Dollars. An important finding of the GSR and other studies is the correlation between ambitious targets and policies and resulting economic benefits from developing and deploying renewables. China, the European Union and Brazil, all with ambitious policies – at least for some sources and/or technologies – have the highest employment figures in renewable energy. In the European Union, Germany alone has more than 370,000 employees in renewables, more than much bigger India, and nearly as much as the USA, where framework conditions are frequently changing und thus creating investment uncertainties.

4   EUROPEAN UNION: FRONTRUNNERS ON THEIR WAY

The European Union can rightly claim to be a body where support for renewables and successful deployment has started early. First support frameworks were established already in the 1980s.

---

[11]The most prominent and frequently quoted study is the World Energy Outlook (WEO) of the International Energy Agency (IEA), which – in 2011 and in 2012 again – suggested that maximum tolerable carbon concentration might soon be exceeded without ambitious and decisive action.

[12]The figures here are from the 2011 edition of the report, published in June 2012. The next edition – to be published in June 2013 – will probably confirm these effects.

As a result of early community policies but also – or even more so – due to ambitious national decisions, the EU counts among its members several frontrunner countries for renewable energy development. They have successfully created support frameworks for market introduction and penetration of sustainable renewable energy.

Denmark is the pioneer country for wind energy development. Back in the 1970s, support for renewable energy – particularly for windpower – was established. From the mid-eighties to 2000, feed-in tariffs supported wind power development and made the country a frontrunner in industry development and turbine installation figures. The share of renewable energy in Denmark's electricity supply is one of the highest in Europe. Two of the world's largest windpower manufacturers, *Vestas* and *Siemens* (formerly *Bonus*), are of Danish origin. Denmark is aiming for 100% renewable energy in electricity and heating and cooling by 2050, with windpower as the major source.

Spain is another pioneer of windpower in Europe, with feed-in tariffs in place from the mid-nineties onwards. In 2011, more than 15% of Spain's electricity was sourced from windpower. Spain is the world's number 4 in installed wind capacity, after China, USA and Germany, with *Gamesa* being one of the world's largest manufactures of wind turbines (GWEC 2013, WWEA 2012). Number 5 is India, with strong growth envisaged for the coming years, whereas in Spain – due to a policy overhaul – new wind power projects (like all other new renewable energy projects) are no longer supported at present (REVE 2012). Spain therefore risks losing the momentum which so successfully helped to increase renewables capacity in recent years.

Germany is widely considered to be the role model for successful renewable energy support policies, in particular in the electricity sector. Driven by a feed-in tariff law[13], the share of renewables increased more than threefold from the mid-1990s to 2012, now accounting for more than 23% of the country's electricity consumption[14], with wind, biomass, and more recently solar PV being the fastest growing technologies. German manufacturers – such as *Enercon* (windpower) and *SolarWorld* (PV) – count among the most important players on global renewable energy markets. Based on the existing framework conditions and the policy decisions to completely phase out nuclear power by 2022 and to make renewables the dominant sources of energy (*Energiewende*) there is reason to assume that the country can maintain its pioneering role in renewables development. And Germany is likely to become one of the major proponents and organizers of a system transformation towards a renewables based energy system.

There are other countries in Europe with high shares of renewable energy. Sweden, Finland, Austria and Latvia have high shares of renewables in their energy mix. In contrast to Denmark, Spain, Germany, and also Portugal where there is a focus on windpower development, these countries have concentrated their efforts on various forms of biomass use, for electricity and for heating and cooling. The progress of renewables achieved so far, in the EU as a whole as well as in the 27 Member States, will be analysed in in this book. It will be shown that there is a good chance for Europe to reach or even exceed the target of 20% renewables by 2020, if stable and reliable framework conditions and enabling policy frameworks remain in place and are further developed.

---

[13] An English text can be found at http://www.erneuerbare-energien.de/fileadmin/ee-import/files/english/pdf/application/pdf/eeg_2012_en_bf.pdf. This version entered into force as of January 2012. More recent amendments regarding PV support are not yet available in English. A German version can be found here http://www.erneuerbare-energien.de/fileadmin/ee-import/files/pdfs/allgemein/application/pdf/eeg_konsol_fassung_120629_bf.pdf

[14] From the webpage of the German Ministry for the Environment, Nature Conservation and Nuclear Safety (BMU), which is in charge of renewable energy policies in Germany: http://www.bmu.de/en/service/publications/downloads/details/artikel/development-of-renewable-energy-sources-in-germany-in-2011-graphics-and-tables/?tx_ttnews%5BbackPid%5D=937&cHash=a8ded3fa6a841588f99adbcd59c0cf4c

5 SCENARIOS AND VISIONS: TOWARDS A POST-2020 FRAMEWORK

Long before the European climate and energy framework for 2020 was agreed, enacted and implemented, there had been policies and (indicative) targets for promoting renewable energies (Chapter 2 of this book). For years the European Parliament and renewable energy associations had been asking for a stable framework for all sectors of renewables, with a focus on an ambitious mid-term policy target. In 2004, EREC[15] launched a campaign for a binding 20% minimum target for renewables by 2020. The proposal was underpinned by scenario calculations extrapolating growth rates until and beyond 2020.

EREC's industry scenarios could build on modelling done by scientists on behalf of the European Commission, which suggested that a 20%-target would be ambitious but realistic. Without the political process leading to an agreement of the heads of state and government in 2007, the target would not have become a legal reality, but without the scenario work the agreement would have been much more difficult, if not impossible.

The agreement of 2007 on 2020 targets and related legislation was a landmark milestone for renewables development in Europe. Now discussion and decision about the next step are due. Once more, potential pathways towards a clean, sustainable and secure energy mix for Europe are explored. In March 2011, the European Commission published a *Roadmap for moving to a Low-carbon Economy in 2050* (European Commission 2011a), exploring options for achieving 80–95% greenhouse gas reduction by 2050 and 30% by 2020. In December 2011, an *Energy Roadmap 2050* (European Commission 2011f) was presented to the public, which explored two business-as-usual scenarios and five decarbonization scenarios, all meeting the 2050 greenhouse gas redaction goals of minus 80–95%. Despite ongoing discussions about underlying assumptions and intransparent methodology, the *Energy Roadmap 2050* has become an important reference document showing – in line with other scenarios presented by scientists and stakeholders – that decarbonizing Europe's energy supply necessarily implies very high shares of renewables (Chapter 9).

There seems to be a broad agreement including the European Parliament's majority, the European Council and most of the Member States that 2050-scenarios with very high shares of renewables are a significant input to the debate about Europe's future security of energy supply. However, 2050 is too far away to serve as an effective trigger for a dedicated policy framework. Part of the consensus on the Roadmap is founded on the long time perspective – wait and see. Nevertheless, it is certainly worth mentioning that only a few years ago scenarios with more than 80% renewables in the electricity supply would have been criticised – if not ridiculed – as utopian. Meanwhile, reality has proven that renewables are mainstream energy sources, with costs decreasing rapidly, through economies of scale and enabling framework conditions.

Technical feasibility and supportive scenarios alone, however, are not enough to facilitate implementation. In addition, short and mid-term political agreement is needed on ambitious and realistic objectives – and enabling legislation. In 2007, a framework up to 2020 was crafted and agreed. Now it is time to prepare a 2030 framework, based on targets which reinforce the momentum created by the 2020 agreement. This is why EREC has launched a proposal for a new legally binding renewables target to follow-up on the 2020-target of 20% – aiming at a share of 45% renewables by 2030 (EREC 2011).

6 FACILITATING THE PARADIGM SHIFT TOWARDS RENEWABLE ENERGY

The political process towards a climate and energy policy framework for Europe beyond 2020 has started and reached a certain degree of maturity. The European Council of December 2012 has

---

[15]The European Renewable Energy Council (EREC) www.erec.org is the umbrella organization of Europe's renewable energy industry, trade and research associations.

mandated the European Commission to prepare a post-2020 framework including 2030-targets (Council 2012b). Like in 2007, as a next milestone, a renewables target for 2030 will need to be part of an integrated climate and energy package. This has been underlined by some Member States and – providing a number of strong arguments – by the European renewable energy sector (EREC 2013b). Some Member States and some stakeholders are advocating a mere greenhouse gas reduction target and some of them are asking to replace all specific support for renewable energy and energy efficiency by the emissions trading system, if anything at all. A majority of MEPs, industry and other stakeholders seem to have understood that there needs to be a specific renewables target to provide a stable framework for investors and industry. It is too early, however, to forecast the outcome of the political process for a post-2020 framework. Whatever the outcome will be, the decision about the 2030-framework will have significant impact on Europe's energy future.

Chapter 11 of this book analyses the present debate and outlines the need for an ambitious and binding renewables target for 2030 combined with a reliable policy framework. The new framework will have to build on existing European legislation and national support systems. It will facilitate further development of synergies from beginning cooperation and interconnection. The post-2020 framework will have to move beyond just increasing the shares of renewables and integrating them into the existing energy system. It will have to facilitate system transformation towards a flexibility driven energy system which needs to be smart enough to integrate dominant shares of renewable energy – variable and dispatchable, centralized and decentralized.

For reasons of transparency and creating a level playing field, external costs of conventional and nuclear energy will have to be fully included in energy prices. Competitive advantages and market distortions favouring conventional and nuclear energy need to be removed. Subsidies for unsustainable energy are incompatible with a sustainable energy future. With a level playing field in place, however, renewables – combined with energy saving and energy efficiency – would be the cheapest energy sources. It is therefore an economic imperative to completely phase out fossil and nuclear energy and to move towards 100% renewables. Early agreement on unambiguous and effectively enabling framework conditions will significantly reduce cost of the necessary transition towards a clean and sustainable energy system.

Section I
The European climate and energy policy framework

CHAPTER 1

Introducing a groundbreaking legislative framework

Rainer Hinrichs-Rahlwes

Europe, the European Union, and in particular some EU Member States are widely held to be frontrunners towards a sustainable energy future. For more than 20 years, Europe has been considered to be actively performing as a pro-active continent for global climate protection and related legal frameworks. The Kyoto Protocol, which entered into force in 2005, was facilitated by perseverance of the European negotiators. Pushing towards an agreement they could build on a set of developed legal frameworks and on ongoing policy support. Increasingly, the enabling role of a strong policy framework, primarily on Member State level but coordinated and agreed on Union level, for renewable energy development became a major backbone of Europe's strong position in favour of effective mitigation measures. Meanwhile, increased deployment of renewable energy has become the main element of greenhouse gas emission reduction in the European Union.

Since policy support for renewable energy started in the European Union more than 25 years ago, wind, solar, hydro, geothermal, and biomass have been moving more and more into the focus of policy debates about the future of our energy supply, about a clean, sustainable and affordable energy mix. Discussions about their benefits for various purposes – including climate protection – started already much earlier. And from the very beginning of this debate, there were a lot of good reasons for supporting the development and deployment of a wide range of renewable energy sources and technologies.

Diversification of resources was a first main driver for supporting development of *alternative* energies, as they were often called in the beginning, in the context of the oil crises of the seventies and eighties. Obviously, renewables were domestic sources of energy. Resources like water for hydropower, wind for driving turbines and sun for producing heat or electricity were locally available everywhere, and they did not have to be imported. Geothermal energy and (sustainably produced) biomass, within the limits of their availability, can also be found and used nearly everywhere, without the need to import raw materials from more or less friendly neighbours or partners and thus contribute to energy indepency. Although experts (particularly those with close links to existing energy companies) voiced doubts about the possibility of scaling up the new technologies, and about their ability to provide relevant contributions to a secure energy supply, there was a broad consensus that it made sense to support renewable energies. Actual competition between renewable sources and incumbent utilities was not yet perceived as a real challenge then.

1.1   HOW TODAY'S POLICY FRAMEWORK WAS DEVELOPED

Section I of this book describes and analyses how support for renewable energy in the European Union started, how it became effective and now is more and more accepted as a mainstream solution for the clean energy mix of today and of tomorrow. The creation and refinement of the climate and energy policy framework of the Union since the beginnings in the 1970s, with a focus on renewable energy, is described and evaluated. The section ends with an assessment of the existing framework and its effectiveness for achieving the 2020 objectives. Section II looks beyond 2020, analysing challenges and policy options for a renewables based energy system towards 2050.

In **Chapter 2, Christine Lins** takes a closer look at the creation and implementation of climate and energy policies in the context of the oil crises and developing climate change framework.

She describes the first EU community programmes, starting with R&D support and demonstration projects in the 1980s and early 1990s, including their part in the development of the EU's climate change strategy of the 1990s in the run-up to the agreement on and ratification of the Kyoto Protocol. She covers the period until 2007, when the Heads of State and Government of the European Union eventually agreed on a climate and energy package to take the EU further on the road towards a sustainable and secure energy future.

She analyses the Green Paper of the European Commission of 1996 (European Commissions 1996) and the resulting White Paper of 1997 (European Commission 1997c), which set a first indicative target for a renewables share in the European Union – 12% by 2010. It obliged Member States to set their own indicative national targets and develop appropriate policies for target achievement. The White Paper outlined a *Campaign for Take-off*, which – between 1999 and 2003 – successfully boosted growth of renewables in the EU, particularly onshore wind, solar PV and different biomass technologies.

As a next step she describes the first important pieces of legislation for renewables on EU-level, the Renewables Electricity Directive of 2001 (Electricity Directive 2001) and the Biofuels Directive of 2003 (Biofuels Directive 2003). And she also takes a look at the Energy Performance of Buildings Directive of 2002 (EPBD 2002), which should have some positive impact on renewables development in the heating and cooling sector.

Chapter 3 describes and analyses the process, which led to the political agreement on major elements of a Climate and Energy Package of the European Union. Beginning with discussions before the compromise was reached during the European Council of March 2007, the chapter sheds some light on the development of the eventual consensus and the main drivers which made this possible.

The legal documents (Emissions Trading Directive, Effort Sharing Decision, CCS-Directive, and the Renewable Energies Directive) for implementing the Council agreement are described in their – partly controversial – development. The final legislative texts are assessed on the background of the process and the targets to be achieved. Evaluation is provided for the main elements of the framework, particularly with regard to the binding targets for the European Union and for each Member State, the controversy about the role of national support systems, and the importance of the National Renewable Energy Action Plans.

The policy and regulatory framework for renewable energy is evaluated with regards to all three sectors (electricity, heating & cooling, and transport) and with a focus on important aspects to be addressed, such as the removal of administrative barriers, and the integration of sustainability requirements with a focus on biofuels for transport. Finally, analysis is provided of the potential impacts of the foreseen evaluations of the directives on the stability of the framework and on investors' confidence.

1.2   HOW THE FRAMEWORK WAS REFINED AND IMPLEMENTED

In **Chapter 4**, **Jan Geiss** takes a closer look on the legal and practical implementation of the framework, with a focus on the transposition of the Renewables Directive into national legislation in the 27 Member States of Union. He analyses the National Action Plans and their potential for target achievement. And he analyses the first interim reports – and some more recent statistical data, which became available later – of the Member States about their target reaching by the end of 2010, which were due in 2011. Including analysis of policy development since the framework entered into force, he provides a first assessment of the expected development of the different elements towards 2020.

In **Chapter 5**, **Thorsten Müller** and **Markus Kahles** take a closer look at national support schemes for renewable energy from the point of view of basic European legal principles. They analyse potential conflicts between these principles and national support mechanisms for renewable energy, most of which are targeted on national production and consumption only, thus excluding (potential) market players from other Member States. The authors provide criteria for potential conflicts between national frameworks and EU basic law, as well as for compliance of national support systems with primary European law. Finally, they provide an outlook about how this debate could further develop.

In **Chapter 6**, again written by **Thorsten Müller** and **Markus Kahles**, in depth analysis is provided for the increasingly relevant discussion about the possibility and/or need for more cooperation between Member States with regard to their support policies for renewable energy, with a focus on the legal problems that could arise within the framework of the Treaty of the European Union. In addition to the policy analysis of Chapter 3, Chapter 6 focuses on legal implications of existing national policies, their compatibility with internal market rules, and on implications of the ruling of the European Court of Justice on the German feed-in system.

The authors start with analysing existing support systems, in particular their main types (feed-in tariffs, feed-in premium and quota obligations with and without tradable green certificates). The authors address differences and similarities of the various systems. They discuss impacts of the proposed solution via harmonisation of support mechanisms throughout the EU. They suggest that potential legal problems of competing national policies could be overcome by more cooperation among Members States and resulting convergence of the frameworks. The cooperation mechanisms of the Renewables Directive are analysed with regard to their contribution to solving conflicts between national policies and basic principles enshrined in the Treaty for the European Union.

Eventually, **Chapter 7** deals with the legislative packages, which – between 1996 and 2009 – were discussed and decided in the European Union in order to facilitate energy market liberalization and market functioning. The chapter focuses on the main elements of the three single market packages, looking at implementation and remaining shortcomings. Eventually, the chapter assesses the impacts of remaining market failures within and between Member States, resulting partly from incomplete implementation of existing EU legislation and partly from shortcomings of the legislation itself.

1.3   WHICH QUESTIONS ARE TO BE ADDRESSED AND ANSWERED

Section I should provide an understanding that a solid ground has been laid for ambitious and stable growth of renewable energy sources in the EU – for energy supply security, economic growth and greenhouse gas reduction. It should provide an understanding that the policy framework was not set up overnight, but that it was the result of an interesting, partly controversial and still ongoing debate about the future of Europe's energy supply.

So far, only a few stakeholders and Member States have pronounced doubts about the possibility and the necessity to fully and ambitiously implement the 2020-targets and framework. Despite the economic and financial crisis, only a few countries have revised their support frameworks with the intention to limit future growth of renewables – but there are some (see Chapter 4 of this book) becoming more outspoken now. And there are those Member States, which have always preferred other *low carbon technologies*, in particular nuclear power and "clean" coal. They are trying to argue about allegedly too high costs of implementation for the 2020-targets. This aspect is addressed in the Outlook-Chapter at the end of this book.

Talk about Carbon Capture and Storage (CCS) as a relevant contribution to greenhouse gas mitigation is fading, no new projects are being developed and even less implemented. It seems that expectations about economic viability (and public acceptance) are approaching those which have long since doubted, if CCS will ever – in the next decades – become economically viable and thus contribute to greenhouse gas emissions mitigation. This is also addressed in the Outlook-Chapter by looking closer at the most recent public consultation about the future of CCS.

In contrast to fading confidence in CCS and in contrast to some politicians' expectations, support for nuclear as an allegedly clean energy source has not really faded (after a short intermission) in some Member States and in parts of the European Commission – at least on the level of vocal commitments. Some countries and their representatives are continuing to advocate nuclear power as an important *low carbon energy source* – despite the Fukushima disaster and the resulting loss of public acceptance and despite the fact that the few new nuclear power plants planned in the EU are either lagging severely behind their schedules and/or dramatically exceeding original cost calculations. Recent UK legislative proposals to support new nuclear power plants by means of *contracts for difference*, granting a *low carbon floor price* (basically copying feed-in models for renewables) for 40 years are only the last examples of how old thinking is prevailing – most likely resulting in burning money, instead of providing affordable and sustainable energy from renewables.

Section I should show that a good framework for renewables is in place for the period up to 2020, which has the potential to facilitate further growth of renewables and resulting economic benefits.

The most efficient and cost effective way to reap all the benefits of the existing framework for greenhouse gas reduction and renewables as the main sources, which are available today, could be paved by an agreement about a new and ambitious mid-term target for the renewable energy share in 2030, and the policy framework needed for implementation. This – and a vision towards 2050 – is what Section II of this book is going to address.

CHAPTER 2

From cradle to adult life: European climate and energy policies until 2007

Christine Lins

2.1   THE BEGINNINGS OF COMMUNITY SUPPORT FOR RENEWABLE ENERGY

The energy sector of the EU is based mainly on fossil fuels, almost two-thirds of which are imported. Today, the EU imports more than 80% of the oil and more than 60% of the gas it consumes. If the current trends continue, import levels will reach more than 70% of the EU overall energy needs by 2030 (European Commission 2011e).

Renewable energy sources (RES) started to be developed when the oil crises of the 1970's made everyone aware of the fact that fossil resources would run out one day. However, controversy remained and remains until today about when exactly that will happen. In its World Energy Outlook 2010 (WEO 2010), however, even the International Energy Agency acknowledged that conventional crude oil production has already peaked in 2006. As far as "peak coal" is concerned, controversial opinions see it happen in the immediate future or within the next 200 years.

What recent policy discussions clearly show, however, is the need to reduce CO_2 emissions of our energy system, therewith largely favouring renewable energy sources. Nevertheless, the European Union realized at a rather early stage that renewable energy sources require strong, continued and smart political commitment.

2.2   THE FIRST COMMUNITY SUPPORT PROGRAMMES[1]

Development of renewable energy has for some time been a central aim of Community energy policy, and as early as 1986 the Council (Council 1986) listed the promotion of renewable energy sources among its energy objectives. Significant technological progress has been achieved since then thanks to the various Community research, technology development and demonstration programmes such as JOULE, THERMIE, INCO and FAIR (for a list of major support programmes and their range see Table 2.1), which not only helped creating a European renewable energy industry in all sectors of renewables but also achieving a world-wide leading position of that industry. In the early 1990s it became clear that, in addition to the efforts made over more than thirty years to develop renewable energy technologies through community research, demonstration and innovation programmes, a policy framework combining legislative and support measures was necessary to foster renewable energy market penetration.

With the ALTENER programme, the Council for the first time in 1993 (Council 1993) adopted a specific financial instrument for renewables promotion. The European Parliament for its part has constantly underlined the role of renewable energy sources and strongly advocated a community action plan to advance them.

The community policy framework established indicative mid-term targets for 2010 both at an overall and a sectoral level. The first time the European Community proposed a target for

---

[1]Taken from section 1.1.3 *"The Need for a Community Strategy of Communication"* from European Commission 1998.

Table 2.1.   Community support programmes.

Name of programme	Timing	Financial volume	Focus
Research & technology development framework programmes	since the early 1970's		
SAVE programme	1992–2002	>100 M€	energy efficiency
ALTENER programme	1993–2002	>120 M€	renewable energy
Intelligent Energy – Europe programme	2003–2006	>250 M€	continues the actions under SAVE and ALTENER
Intelligent Energy – Europe programme*	2007–2013	>730 M€	embedded within the Competitiveness & Innovation Programme (CIP)

*http://ec.europa.eu/energy/intelligent/.

renewable energy was in the ALTENER programme in 1993: the objective was to double the contribution of renewable energy to gross domestic consumption from 4% in 1991 to 8% by 2005.

2.3   THE PROMOTION OF RENEWABLE ENERGY AS CENTRAL PILLAR IN THE FIGHT AGAINST CLIMATE CHANGE

Five years after the Rio Conference that took place in 1992, climate change was again at the centre of international debate in view of the upcoming "Third Conference of the Parties to the United Nations Framework Convention on Climate Change" to be held in Kyoto in December 1997. The European Union recognised the urgent need to tackle the climate change issue. It also adopted a negotiating position of a 15% greenhouse gas emissions reduction target for industrialised countries by the year 2010 from the 1990 level. To facilitate the Member States achieving this objective, the Commission, in its communication on the *Energy Dimension of Climate Change* (European Commission 1997a) identified a series of energy actions – including a prominent role for renewables.

The Council of Ministers endorsed this when inviting the Commission to prepare an action programme and present a strategy for renewable energy. In preparation for the international climate change conference in Kyoto, the Commission confirmed the technical feasibility and economic manageability of the Union's negotiating mandate. In its Communication *"Climate Change – The EU Approach to Kyoto"* (European Commission 1997b), the Commission analysed the consequences of reducing CO_2 emissions significantly, including the implications for the energy sector. In order to achieve such a reduction, the Union required major energy policy decisions, focusing on reducing energy and carbon intensity. Accelerating the penetration of renewable energy sources was considered very important for reducing carbon intensity and hence CO_2 emissions, whatever the precise outcome of the Kyoto Conference.

2.4   THE BREAK-THROUGH: THE 1997 WHITE PAPER BEING THE FIRST LEGISLATIVE ELEMENT ON RENEWABLE ENERGY IN THE EU

The 1997 Communication from the Commission *"Energy for the future: renewable energy sources – White Paper for a Community Strategy and Action Plan"* (European Commission 1997c) followed on from the discussion stimulated by the Green Paper (European Commission 1996) on the same topic published by the Commission in November 1996.

In the Green Paper on Renewables the Commission sought views on the setting of an indicative objective of 12% for the contribution by renewable sources of energy to the European Union's gross inland energy consumption by 2010. The overwhelmingly positive response received during the consultation process confirmed the Commission's view at that time that an indicative target is a good policy tool, giving a clear political signal and impetus to action.

The White Paper therefore set an indicative target of doubling the share of renewable energy of the European Union's overall gross internal energy consumption from 6% to 12% by 2010 which was considered an ambitious but realistic objective.

Given the overall importance of significantly increasing the share of RES in the Union, this indicative objective was considered as an important minimum objective to be maintained, whatever the precise binding commitments for CO_2 emission reduction would finally be. The importance of monitoring progress and maintaining the option of reviewing this strategic goal if necessary was underlined.

The calculations of increase of RES needed to meet the indicative objective of a 12% share in the Union's energy mix by 2010 were based on the projected energy use in the pre-Kyoto scenario. It was considered likely that the projected overall energy consumption in the EU 15 might decrease by 2010, if the necessary energy saving measures were taken post Kyoto. At the same time, the enlargement of the Union to new Member States where RES were almost non-existent would require an even greater overall increase. It was therefore considered at this stage that the 12% overall objective could be refined further and furthermore it was clearly emphasized that this overall objective was a political and not a legally binding tool.

The *Community Strategy and Action Plan* should be seen as an integrated whole which had to be further developed and implemented in close cooperation between the Member States and the Commission. The challenge faced required a concerted and coordinated effort by the various players over time. Measures should be taken at the appropriate level according to the subsidiarity principle within the coordinated framework provided by this Strategy and Action Plan. It would be incorrect and unrealistic to assume that actions needed only to be taken at Community level. From the beginning, it was pointed out that Member States have a key role to play in taking the responsibility to promote renewable energy sources, through national action plans, to introduce the measures necessary to promote a significant increase in renewables penetration, and to implement this Strategy and Action Plan in order to achieve the national and European objectives.

It was agreed that legislative action at EU level would only be taken if measures at national level were insufficient or inappropriate and when/if harmonisation was required across the EU.

Main features of the proposed action plan aimed at providing fair market opportunities for renewable energy sources without excessive financial burdens. For this purpose, a list of priority measures was drawn up, including:

- non-discriminatory access to the electricity market;
- fiscal and financial measures;
- new initiatives regarding bio-energy for transport, heat and electricity and, in particular, specific measures to increase the market share of Biofuels, promote the use of biogas and develop markets for solid biomass;
- the promotion of the use of renewable energy sources (such as solar energy) in the construction industry, both in retrofitting and for new buildings.

It was also clear that the reaching of this objective would require a major input from the Member States, which must promote wider use of renewable energy sources as far as their potential allows.

Member States were encouraged in their efforts to set their own targets in order to

- make greater use of the potential available;
- help further cut CO_2 levels;
- reduce energy dependence;
- develop the national industry;
- create jobs.

It was estimated that substantial investment in the order of ECU[2] 95 billion for the period from 1997 to 2010 were needed to achieve this overall goal. Greater use of renewable energy sources was expected to provide substantial economic benefits, in particular major export opportunities in view of the European Union's capacity to supply equipment as well as technical and financial services.

Estimates also pointed to:

- the creation of 500,000 to 900,000 jobs;
- an annual saving of fuel costs of ECU 3 billion from 2010;
- the reduction of fuel imports by 17.4%;
- the reduction of CO_2 emissions by 402 million tonnes a year by 2010.

Before then, little importance was attached to renewable energy sources in Community policies, programmes and the budget. The action plan therefore aimed to raise the awareness of those responsible for the various programmes and to give greater prominence to renewable energy sources in the Union's various policies, such as:

- environment;
- employment;
- competition and State aid;
- technological research and development
- regional policy;
- the Common Agricultural Policy and rural development;
- external relations, in particular through the PHARE, TACIS and MEDA programmes, etc.

To achieve the objective set in the White Paper, cooperation between the Member States was to be increased. Support measures were also provided for, in particular under the ALTENER programme, to achieve targeted action, to inform consumers, to develop European standards, to improve the position of renewable energy sources on the capital market of the institutional and commercial banks and to create networks (of regions, islands, universities, etc.) in the field of renewable energy sources.

2.5   THE CAMPAIGN FOR TAKE-OFF

Even though renewable energy technologies had by then reached a certain maturity, there were many obstacles to their market penetration. In order to assist a real take-off of renewables for large scale penetration, make progress towards the objective of doubling the EU renewable energy sources share by 2010, and ensure a coordinated approach throughout the Community, the Commission proposed a Campaign for Take-Off (European Commission 1997c, p. 27ff) of renewables. This would be undertaken over a number of years in close cooperation between the Member States and the Commission. The proposed campaign aimed to promote the implementation of large-scale projects in different renewable energy sectors in order to send clear signals for greater use of renewable energy sources.

The aim of this campaign was to boost high-profile projects in various renewable energy sectors. Several key actions were promoted during the campaign:

- the installation of one million photovoltaic systems, with 500,000 for roofs and façades for the EU domestic market (total investment cost: ECU 1.5 billion) and 500,000 for export, in particular to kick-start decentralised electrification in developing countries;
- 10,000 MW of large wind farms;
- 10,000 MW_{th} of biomass installations;

---

[2]ECU: European Currency Unit replaced by EURO on 1 January 1999, for further information see http://en.wikipedia.org/wiki/European_Currency_Unit

- integration of renewable energies in 100 small communities, regions, conurbations, islands, etc. as a pilot scheme.

It was agreed that the implementation of the Strategy and Action Plan set out in the White Paper would be closely followed, under the ALTENER programme, to measure the progress made in the penetration of renewable energy sources.

The Campaign for Take-Off (CTO) was the first European promotion campaign for RES launched in 1999 and concluded in December 2003. It was an essential part of the strategy outlined in the *"White Paper for a Community Strategy and Action Plan on renewable energy sources"*, as it was designed to kick-start the implementation of this legislative document.

The key sectors identified by the CTO corresponded to technologies which were mature at that time and therefore were considered crucial for achieving the White Paper's RES goal but which needed an initial stimulus to accelerate and substantially improve their market penetration, thereby developing economies of scale and, consequently, reducing costs.

The Campaign for Take-Off aimed to facilitate the success of the strategy as a whole by stimulating the necessary trend towards increased private investment in renewables in a visible manner, with an emphasis on near-market technologies – solar, wind and biomass. The CTO was expected to have reached its goals by 2003, i.e. to set out a framework for action to highlight investment opportunities and attract the necessary private funding which is expected to make up the lion's share of the capital required.

The CTO also sought to encourage public spending to focus on the key sectors, which is expected, in the process, to trigger private investment as a result. Lastly, the CTO was known for its communication strength by launching several types of partnerships and promotional activities (impact assessment of the CTO in EREC 2004).

2.6   LEGISLATION FOR RENEWABLE ENERGY USE IN THE ELECTRICITY AND TRANSPORT SECTOR AS WELL AS FOR BUILDINGS

Further to the Commission's White Paper, a European legislative framework to promote renewable energy was established in the electricity and the transport sector, with two specific EC directives that established growth targets for renewable energy in these respective areas, both at Community and national level, as well as a series of specific measures and monitoring schemes.

2.6.1   *The RES-electricity Directive*

The *Directive on the Promotion of Electricity produced from Renewable Energy Sources* (Electricity Directive 2001), adopted in September 2001, which enabled, for the first time, concrete national indicative targets to be agreed with the Member States to sustain a substantial increase of renewables electricity, passing from 14% in the year 2000 to 22.1% in EU-15 (21% in EU-25) by 2010 (for details see Table 2.2). The directive required Member States to take appropriate steps to encourage greater consumption of electricity produced from RES by setting and achieving annual national indicative targets consistent with the directive and national Kyoto commitments.

It constituted an important milestone in shaping the regulatory framework for RES-E generation in the EU. The RES-electricity Directive contained the following measures:

- Quantified national targets for consumption of electricity from renewable sources of energy; Member States had to publish national RES electricity targets by October 2002, independently of the support system or scheme into force.
- National support schemes plus, if necessary, a harmonised support system.
- Simplification of national administrative procedures for authorisation.
- Guaranteed access to transmission and distribution of electricity from renewable energy sources.

Table 2.2.   National indicative targets as set out in the Directive on the promotion of electricity from renewable energy sources.

	Share of RES-Electricity in %		Produced RES-Electricity in TWh	
	1997	2010	1997	2010
Austria	72.7	78.1	51.5	55.3
Belgium	1.1	6.0	1.2	6.3
Denmark	8.7	29.0	3.9	12.9
Finland	24.7	35.0	23.2	33.7
France	15.0	21.0	80.6	112.9
Germany	4.5	12.5	27.5	76.4
Greece	8.6	20.1	6.2	14.5
Ireland	3.6	13.2	1.2	4.5
Italy	16.0	25.0	57.3	89.6
Luxembourg	2.1	5.7	0.2	0.5
Netherlands	3.5	12.0	4.6	15.9
Portugal	38.5	45.6	23.9	28.3
Spain	19.9	29.4	51.8	76.6
Sweden	49.1	60.0	79.8	97.5
United Kingdom	1.7	10.0	8.5	50.0
European Union	13.9	22.1	424.5	674.9

The RES-E Directive provided for a broad definition of renewable energy. It included hydropower (large and small), biomass (solids, biofuels, landfill gas, sewage treatment plant gas and biogas), wind, solar (PV, heat, thermal electric), geothermal, wave and tidal energy. General waste incineration was excluded but the biodegradable fraction of waste was considered as renewable. The contentious category biodegradable part of waste incineration 'as long as the waste hierarchy is respected' was retained.

Furthermore, large hydropower (more than 10 MW) was also included. It was tacitly agreed that large hydro would count for meeting the targets but would not be eligible for support measures.

Back then and still today, Member States operate different mechanisms of support for RES-E at the national level, such as investment aid, tax exemptions or reductions, tax refunds and aid supporting the price paid to the producer (direct price support), the latter being in most Member States the principal promotion tool for RES-E.

It is important to note that the RES-E Directive did not prescribe a harmonised European support scheme; it was stipulated explicitly that the national systems would continue to exist. The Member States were obliged to fulfil their own clearly specified national targets, which vary greatly. These targets were agreed by a burden-sharing process. And at the same time, the principles providing for these national targets for consumption of electricity from renewable sources of energy were defined at the Community level.

However, as indicated in the Commission staff working document accompanying the directive proposal, although the Commission considered it to be inappropriate at that moment to harmonise European support schemes, it noted that *"this does not, of course, preclude Member States from taking measures to harmonise support schemes from the 'bottom-up', such as by linking up their green certificate regimes or developing common feed in schemes. A reduction in the number of different support schemes could generate substantial economies of scale, simplify the regulatory environment and increase transparency for investors, and hence allow a more cost-effective achievement of the renewable targets"* (European Commission 2008, p. 14).

The years following the publication of the RES-E Directive were accompanied by sometimes very emotional discussions on the need and value to harmonise support schemes or not; various

Table 2.3.   Biofuels development and targets 1995–2010.

	1995 Eurostat Mtoe	2000 Eurostat Mtoe	Annual growth rate 1995–2000 %	Directive target 2010 Mtoe	Annual growth rate needed 2001–2010 %
Biofuels	0.27	0.68	20.2	17.0	38.0
Gasoline and oil demand	237.7	256.6	1.5	295.8	1.6
Biofuels' share	0.1	0.26		5.75	

project looked into this question such as the Futures-e project[3] just to name one of the most recent ones.

The directive just stipulated that the Commission would make an analysis of and an investigation on each national system by 2005 which led to the conclusion that harmonisation was premature and would lead to investor insecurity and therefore be counterproductive to RES deployment in the EU.

The *Directive of the European Parliament and of the Council on the promotion of the use of energy from renewable sources amending and subsequently repealing Directives 2001/77/EC and 2003/30/EC* (Renewables Directive 2009), contained in the 2008 Climate and Energy Package[4], is building on the successful Electricity Directive 2001 and the Biofuels Directive 2003. In addition, it is closing the gap by providing legislation on the use of renewable energy in the heating and cooling sector. See Chapter 3 for further reference to the Climate and Energy Package.

2.6.2   *Legislation on biofuels*

The Directive on the promotion of biofuels' use for transport (Biofuels Directive 2003), adopted in May 2003, which set a target for the increase of biofuels consumption (2% by 2005 and 5.75% by 2010) as a proportion of total petrol and diesel consumption, compared to 0.6% at the moment of the adoption of the directive (for details see Table 2.3). Biofuels include bioethanol, biodiesel and any fuel for transport produced from renewable energy sources. National governments should, according to the directive, introduce measures to promote the production of biofuels in their territory.

In October 2003, the Council adopted a new fiscal directive on the taxation of energy products (Energy Taxation Directive 2003). The directive widens the scope of the EU's minimum rate system for energy products, previously limited to mineral oils, to all energy products including coal, natural gas and electricity, thereby also contributing to the promotion of biofuels. It was stipulated that energy products were taxed only when used as a fuel or for heating. Amongst other provisions, the directive allowed Member States to exempt RES, including biofuels.

2.6.3   *Directive on the promotion of energy performance of buildings*

In addition to the electricity and the biofuels directives, another piece of legislation was developed focusing on the promotion of energy performance of buildings (EPBD 2002). Given the low turnover rate of buildings, with a lifetime of fifty to more than one-hundred years, it is clear that the largest potential for improving energy performance in the short and medium term could be found in the existing building stock. The directive laid down a framework leading to increased co-ordination of legislation in this field between Member States. It was decided, however, that the

[3]http://www.futures-e.org/
[4]http://ec.europa.eu/clima/policies/package/index_en.htm (viewed on 29 April 2013).

practical application of the framework would remain primarily the responsibility of the individual Member States. The directive covered five main elements:

- Establishment of a general framework of a common methodology for calculating the integrated energy performance of buildings;
- Application of minimum standards to the energy performance to new buildings and to certain existing buildings when they are renovated;
- Certification schemes for new and existing buildings on the basis of the above standards and public display of energy performance certificates and recommended indoor temperatures and other relevant climatic factors in public buildings and buildings frequented by the public;
- Specific inspection and assessment of boilers and heating/cooling installations;
- National Policy Measures following the EU legislation.

Once energy performance certificates for buildings would be widely used (which is nowadays the case), the price of buildings (rented or sold) will start reflecting their energy performance.

The directive proposed the following cost-effective measures to improve the energy performance of buildings within the EU with the objective to facilitate convergence of building standards towards those Member States which already have ambitious levels:

- Methodology for integrated buildings energy performance standards;
- Application of standards on new and existing buildings;
- Certification schemes for all buildings;
- Inspection & assessment of boilers/heating and cooling installations.

2.7   RENEWABLES HEATING & COOLING: THE MISSING LINK

Although the European Union was a frontrunner in RES-electricity and biofuels promotion, there was a clear legislative gap in the field of promotion of heating and cooling with renewable energy.

Over 40% of the primary energy consumption in Europe is used for heating buildings, for domestic hot water production and for heating in industrial processes. Heat is the largest consumer of primary energy, being greater than electricity or transport. The markets for renewable heating sources (biomass, solar thermal, geothermal) therefore have a substantial potential for growth, and could replace substantial amounts of the fossil fuels and electricity, which are currently used for heating purposes.

However, both at EU level and in most Member States, renewable heating by then had received less political attention than renewable electricity, largely because renewable heating is not traded using European-wide networks. Another reason may be that renewable heating is sold mainly by SMEs, which did not yet have a strong identity in the EU energy markets. The introduction of support schemes was considered to be more complicated for renewable heating than for renewable electricity, because there is not a single European market for renewable heating, the market is not regulated by a single regulator, and the monitoring of sales is not co-ordinated by a single entity.

Nevertheless, renewable heating was a key component of the 1997 White Paper on renewable energy sources, and therefore voices both from the European Parliament as well as the renewable energy industry (EREC 2005) were raised that it should form a major component of any coherent strategy to develop renewable energy sources. As concluded in the *"Communication from the Commission on the implementation of the Community Strategy and Action Plan on Renewable Energy Sources (1998–2000)"* (European Commission 2001), *"the definition of individual RES strategies and objectives by Member States, as called for in the proposal for a Directive on electricity from RES. However, the definitions of these objectives should address not only electricity, but also the heating, cooling and transport sectors"* (page 29).

This call ultimately found a response during the discussions on the EU's Energy and Climate Package and measures for accelerating growth of RE in the heating and cooling sector have become an important part of the Renewables Directive (details in Chapter 3 of this book).

CHAPTER 3

The European climate and energy package for 2020

Rainer Hinrichs-Rahlwes

3.1 INTRODUCTION

In the international policy arena, particularly in climate conferences and negotiations about targets and policy frameworks for climate protection, the European Union (EU) had long since been considered to be a global leader. The delegations of the European Union and of the EU's Member States had been strong supporters and advocates of the Kyoto Protocol. And it can be reasonably assumed that without the proactive role of EU negotiators and governments, based on broad civil society support, the protocol would never have been agreed on and eventually been ratified.

However, pushing for an international framework and ratifying it, when the minimum consensus of the signatories has been found and written down, is only one side of the coin, which was (and still is) difficult enough to design and implement. The other side – unconditional implementation of policies and measures assessed to be necessary for climate change mitigation and useful for a secure and future oriented energy supply – is even more complex and more difficult to be designed. And it had not been seriously considered on a broader scale for many years. The traditional way of thinking (insisting on implementing only policies which had been broadly or even globally agreed) is based on the misconception that implementing climate protection measures and green energy options is primarily a burden – and not an asset or a benefit. This was and still is a key factor for slow progress and inaction in global climate protection policies.

The European Union's decision to set up a climate and energy package, basically agreed on at the European Council[1] in March 2007, could therefore be seen as a significant and positive deviation from the path of talking and trying to reach minimum consensus instead of designing and implementing necessary action. Despite some remaining critical aspects, the agreement of the EU's Heads of State and Government and the following detailed implementation in a number of directives[2] is an outstanding global landmark for sustainable climate and energy policies – notwithstanding the fact that the formal agreement on the legal details was not achieved before the COP14/MOP4 in Poznan (Poland) in December 2008.

The Council agreement and the different elements of the climate and energy package and the resulting legislative acts will be described and analysed in this chapter. The analysis will show that the agreement on a set of 2020-targets for the European Union, underpinned by legally binding targets for each Member State in the Renewables directive, created a significant momentum for the development of sustainable climate and energy policies in the European Union as a whole and even more in some Member States.

---

[1]The European Council, which according to the European Treaty is composed of the Heads of State and Government of all EU Member States, and – according the Lisbon Treaty – should discuss and decide *"the general political directions and priorities"* of the European Union is the appropriate institution for political decisions of high importance. Europe's role in international climate protection policies certainly is such an area of paramount importance.

[2]According to the European Treaty, European directives are a key element of European legislation, which have to be transposed into national legislation by every Member State. Non-compliance is usually followed by infringement procedures launched by the European Commission and can eventually be penalized by the European Court of Justice.

3.2   THE COUNCIL AGREEMENT OF MARCH 2007

In the months before the Heads of State and Government of the European Union met in March 2007, there had been a controversial but constructive discussion about some aspects of the climate and energy agenda of the European Council. Climate and Energy was one of the key topics which German Chancellor Angela Merkel had addressed in her Presidency[3] agenda.

Environmental organisations (NGOs) and some governments had been asking for an EU-wide consensus about ambitious targets and commitments in the run-up towards the climate summit to be held in December 2007 in Poznan. Whereas NGOs were clearly asking for the unilateral commitment of the EU to reduce greenhouse gas emissions by at least 30% (compared to 1990 levels) until 2020, most governments were quite hesitant to agree on what they considered to be an overly ambitious unilateral target. Part of this debate was the further development of the European Emissions Trading Scheme (EU-ETS), particularly the issue of auctioning emission rights instead of freely distributing them via a grandfathering procedure. In addition, the important aspect of implementing greenhouse gas reduction policies in "non-ETS" sectors had to be tackled.

Another debate was about the future of renewable energy development and deployment in Europe. Industry representatives, in particular the Brussels based renewable energy organizations[4] and several Members of the European Parliament had been asking to establish a European target to reach a share of at least 25% renewable energy in the heating and cooling sector in 2020. And they had been pushing for a European directive for the promotion of renewable energy in the heating and cooling sector, following the examples of the Electricity Directive 2001 and the Biofuels Directive 2003. Furthermore, there had been urgent lobbying – lead by EREC, the European Renewable Energy Council, the umbrella organization of European renewables associations – for a legally binding renewable energy target of at least 20% in 2020, underpinned by specific targets for the three sectors – electricity, heating and cooling, and transport.

At the end of the Council Meeting, when the presidency conclusions were presented by Chancellor Merkel, most observers agreed that an ambitious package had been developed, with a particularly strong focus on renewable energy. In the official press declaration, the outcome of the Council meeting regarding the climate protection target was described as follows: *"In difficult negotiations with the EU Member States' Heads of State and Government, Federal Chancellor Merkel achieved a breakthrough towards an ambitious integrated European climate and energy policy. This will give the EU credibility in the pioneering role it will play in international climate protection during the negotiations on a post-Kyoto regime, due to begin in 2007. In this context the European Council agreed to make a voluntary commitment to reduce greenhouse gas emissions by 30 percent as long as other industrialized countries adopt comparable goals. Apart from this commitment, however, it defined an independent goal to cut greenhouse gas emissions by 20 percent by 2020 in relation to 1990 levels"* (Council 2007b).

---

[3]The Presidency of the European Union and consequently the Presidency of the European Council changes every six months in a preset – more or less alphabetical – order. In close cooperation with the outgoing and the following presidency, each new presidency usually presents an agenda including the ongoing issues to be addressed and solved during the six months of the upcoming presidency. Although this agenda has to reflect pending issues and the general consensus of the European Union, the incoming presidency usually includes some points of specific interest. Climate protection und renewable energy were such issues, which the German Presidency had specifically committed to.

[4]The European Renewable Energy Council (EREC, the umbrella organization of the Brussels based European renewable energy associations) and EREC's Members, associations representing specific technologies, such as wind, solar, biomass, geothermal and hydropower, or specific aspects of the renewables sector, such as producers of energy from renewable sources or research organizations, have established a solid base of policy work and advocacy for renewables in Brussels, representing an important counter-weight to the conventional energy lobbyists trying to influence the European Commission, as well as Members of the European Parliament and of the Council.

The climate related paragraphs – under the headline of *"An integrated climate and energy policy"* – underlined some of the key agreements. The presidency conclusions highlighted an agreement on the overall target: (Council 2007a, No. 27) *"The challenges of climate change need to be tackled effectively and urgently. Recent studies on this subject have contributed to a growing awareness and knowledge of the long-term consequences, including the consequences for global economic development, and have stressed the need for decisive and immediate action. The European Council underlines the vital importance of achieving the strategic objective of limiting the global average temperature increase to not more than 2°C above pre-industrial levels".*

After underlining (Council 2007a, No. 29) *"the leading role of the EU in international climate protection"*, the Council *"stresses that international collective action will be critical in driving an effective, efficient and equitable response on the scale required to face climate change challenges. [. . .] All countries should be invited to contribute to the efforts under this framework according to their differentiated responsibilities and respective capabilities".*

In addition to the greenhouse gas reduction targets, the European Council decided on a *"European Energy Action Plan: Energy Policy for Europe"* (Council 2007a, p. 16 ff), in which various elements for a future secure and sustainable energy policy are laid out. In this action plan, the key targets and policies for renewable energy and energy efficiency are summarized, or as the official press communication put it (Council 2007b): *"Given the central role of a sustainable energy policy in achieving climate objectives, and as a milestone on the way to a European energy policy, the European Council adopted a European Energy Action Plan with the three goals of security of supply, efficiency and environmental compatibility. The negotiations centred around the agreement on a binding commitment to increase to 20 percent the proportion of renewable energies in overall energy consumption. This agreement is supplemented by the goal to introduce efficiency measures to cut by 20 percent the total energy consumption predicted for 2020. Linking energy saving with a clear commitment to promoting renewables conveys an important message to energy markets to invest in sustainable and innovative energies".*

The text of the action plan describes in more detail, what the Heads of State and Government agreed on (Council 2007a, No. 7, p. 21): *"The European Council reaffirms the Community's long-term commitment to the EU-wide development of renewable energies beyond 2010, underlines that all types of renewable energies, when used in a cost-efficient way, contribute simultaneously to security of supply, competitiveness and sustainability, and is convinced of the paramount importance of giving a clear signal to industry, investors, innovators and researchers".*

Based on this overall assessment, the Council agreed on *"a binding target of a 20% share of renewable energies in overall EU energy consumption by 2020"* and on *"a 10% binding minimum target to be achieved by all Member States for the share of biofuels in overall EU transport petrol and diesel consumption by 2020, to be introduced in a cost-efficient way".*

The binding character of the 10%-target for biofuels was assessed to be appropriate *"subject to production being sustainable, second-generation biofuels becoming commercially available and the Fuel Quality Directive being amended accordingly to allow for adequate levels of blending".*

In addition to these binding targets for the overall share of renewable energy and for the transport sector, the action plan highlights an important additional agreement: *"From the overall renewables target, differentiated national overall targets should be derived with Member States' full involvement with due regard to a fair and adequate allocation taking account of different national starting points and potentials, including the existing level of renewable energies and energy mix [. . .] leaving it to Member States to decide on national targets for each specific sector of renewable energies (electricity, heating and cooling, biofuels)".* For implementation of these agreements, the European Commission is asked to come forward in 2007 with a proposal *"for a new comprehensive directive on the use of all renewable energy resources"*, consisting of – among other things – overall national targets for each Member State, *"National Action Plans containing sectoral targets and measures to meet them"*, and a sustainability clause for biofuels: *"criteria and provisions to ensure sustainable production and use of bioenergy and to avoid conflicts between different uses of biomass".*

Table 3.1.   The European Union's "Climate and Energy Package" of 2008.

ETS Directive
   Directive of the European Parliament and of the Council amending Directive 2003/87/EC so as to
   improve and extend the greenhouse gas emission allowance trading scheme of the Community

Effort Sharing Decision
   Decision of the European Parliament and of the Council on the effort of Member States to reduce their
   greenhouse gas emissions to meet the Community's greenhouse gas emission reduction commitments
   up to 2020

CCS Directive
   Directive of the European Parliament and of the Council on the geological storage of carbon dioxide
   and amending Council Directive 85/337/EEC, Directives 2000/60/EC, 2001/80/EC, 2004/35/EC,
   2006/12/EC, 2008/1/EC and Regulation (EC) No 1013/2006

Renewable Energy Directive (RED)
   Directive of the European Parliament and of the Council on the promotion of the use of energy from
   renewable sources amending and subsequently repealing Directives 2001/77/EC and 2003/30/EC

3.3   IMPLEMENTING THE COUNCIL AGREEMENT

In the months after the Council meeting, the European Commission started to develop the various pieces of legislation needed to implement the agreement by providing the various documents and legislative proposals asked for by the Heads of State and Government. However, despite the agreement on major elements of Europe's future climate and energy policies, it took more than nine months until – on the 23rd of January 2008 – the Commission eventually presented a proposal for a comprehensive climate and energy package, consisting of four main elements (see Table 3.1) and various accompanying documents, such as impact assessments. It took another year until, in 2009, agreement on all parts of the package was reached between the European Parliament, the European Council and the European Commission so that the legislative acts could enter into force and be transposed into national law in each Member State.

More than three years later, the Climate and Energy Package has entered into force and the transposition into national law has been accomplished – in theory, at least. In reality, there are a number of shortcomings. For example, some Member States have not yet completed the legislative process of transposition, some have not correctly transposed all aspects, and some have submitted documents which do not really match with reality.

Already during the process of elaborating the final documents and agreeing on all the necessary details, it became evident that some governments and their civil servants were trying to modify the agreement of March 2007, or at least interpreting the consensus in various different directions. This was particularly evident, when it came to defining the details of auctioning emission rights in the Emissions Trading Directive, and it was even more obvious in the negotiations about the Renewable Energy Directive. After nearly two years of intensive discussions and negotiations about important details, the European institutions managed to find a first reading agreement on all related dossiers – covering all major political questions – before the end of 2008. Remaining points of political importance were discussed and decided in the European Council on December 10, 2008. The final legal texts were produced in the first few months of 2009.

3.4   THE EMISSIONS TRADING DIRECTIVE

The European Emission Trading System was legally established in 2003 and officially launched in 2005 as a cap and trade system, limiting the amount of greenhouse gas emissions for the participating companies and entities. The overall quantity of carbon dioxide, which participants were allowed to emit, was limited by the directive. Based on an overall cap, individual emission rights for the different participants were defined by the number of allowances they could redeem. The initial distribution of emission rights for each participant was primarily based on different types of grandfathering, partly combined with benchmarks. It was designed and implemented in each Member State of the European Union, after having notified the "National Allocation Plan" to the European Commission[5]. Participants needing more emission rights than allocated to them have to try and purchase them from those who are willing and/or able to sell emission rights, which they had been allocated but do not need themselves. Or they have to reduce emissions, either by investing in cleaner technologies or by reducing operation times of the installation. The idea behind the EU-ETS is to limit the available emission rights and thus reduce greenhouse gas emissions. In reality of the first years of the EU-ETS, however, the amount of emission rights allocated to the participants had obviously been too high. Due to overallocation, carbon prices dropped to less then 1 € per tonne, reducing the directive's efficiency and positive impact on emissions reduction drastically.

This is why the *"Directive of the European Parliament and of the Council amending Directive 2003/87/EC so as to improve and extend the greenhouse gas emission allowance trading system of the Community"* (Emissions Trading Directive 2009) became part of the Climate and Energy Package. The main intention of the amendments was repairing the shortcomings of the original version (Emissions Trading Directive 2003) so that the system would eventually take off and actually deliver relevant emissions reductions.

The amended directive basically changed two important aspects of the ETS. As laid out in the text of the Commission's proposal of January 2008 (Emissions Trading Directive 2008), a more European approach was deemed to be necessary and – in order to prevent over-allocation and windfall profits – auctioning instead of grandfathering should become the rule by 2013. This was highlighted in No. 9 and No. 10, page 5 of the draft proposal: *"The effort to be made by the European economy requires inter alia that the additional efforts from the revised EU-ETS be achieved with the highest possible degree of economic efficiency and on the basis of fully harmonised conditions of allocation within the Community. Auctioning should be the basic principle for allocation, as it is the simplest and most economically efficient system. This will also eliminate windfall profits and put new entrants and higher than average growing economies on the same competitive footing as existing producers . . . 10. Consequently, and taking into account their ability to pass on the increased cost of CO_2, full auctioning should be the rule from 2013 onwards for the power sector . . ."*

The Commission's proposal accepted that there might be some installations where free allocation of allowances could still be necessary, e.g. for new market entrants, but free allocation should no longer be granted by the individual Member States, but based on European standards (No. 14, page 6): *"While there is transitional free allocation to installations, this should be through harmonised Community-wide rules in order to minimise distortions of competition with the Community. These rules should take account of the most greenhouse gas and energy efficient techniques, substitutes, alternative production processes, use of biomass, renewables and greenhouse gas capture and storage. Any such rules must avoid perverse incentives to increase emissions. An increasing proportion of these allowances should be auctioned. While there is transitional free allocation to installations, harmonised Community-wide rules for free allocation should apply for to new entrants carrying out the same activities as those installations. To avoid*

---

[5] In the European legal system *notification* is a key instrument for securing compliance with European law. Member States have to elaborate their respective actions, plans, laws and other forms of implementation and – if European legislation requires – submit these plans to the European Commission. The Commission – after consulting the other Member States where necessary – eventually approves (with or without additional requirements) or refuses to approve the actions of the Member States. This procedure of *notification* applied to the Member States' National Allocation Plans (NAPs), the detailed plans indicating criteria and quantities of allocation of emission rights to companies which have to participate in the EU-ETS.

distorting competition within the internal market, no free allocation should be made in respect of the production of electricity by such new entrants."

As could have been expected, apart from the overall level of reduction, the questions related to auctioning of allowances instead of free allocation and the issues related to the distribution of auctioning revenues were the main points of conflict in the Council working groups, the European Parliament's committees and eventually in the European Council in December 2008, where the remaining conflicts had to be solved and decided.

The Commission's proposal of January 2008 would automatically have increased the ETS-Directive's level of ambition from a 20% reduction by 2020 to 30%, if an international climate protection framework were agreed on. This was completely in line with the European Council's consensus of March 2007, but some Member States had changed their minds until December 2008, so that this mechanism could not be included in the final agreement. As a result, the level of ambition of the directive was linked to an EU-target of reducing greenhouse gas emissions by 20% by 2020. For meeting a 30%-target, a new legislative process is foreseen in Article 28 *("Adjustments applicable upon the approval by the Community of an international agreement on climate change")* of the finally agreed directive (Emissions Trading Directive 2009). *"Within three months of the signature by the Community of an international agreement on climate change leading, by 2020, to mandatory reductions of greenhouse gas emissions exceeding 20% compared to 1990 levels, as reflected in the 30% reduction commitment as endorsed by the European Council of March 2007, the Commission shall submit a report . . .".* Based on this report, legislation shall be proposed to implement the international agreement, including the 30% greenhouse gas reduction in the European Union.

A compromise was also found for the distribution of those allowances which have to be auctioned. The new Article 10 *("Auctioning of allowances")* of the ETS-Directive includes strict rules for distribution. As a rule *"88% of the total quantity of allowances to be auctioned being distributed amongst Member States in shares that are identical to the share of verified emissions under the Community scheme for 2005 or the average of the period from 2005 to 2007, whichever one is the highest, of the Member State concerned."* In order to support economically weaker Member States *"10% of the total quantity of allowances to be auctioned being distributed amongst certain Member States for the purpose of solidarity and growth within the Community . . ."* And finally, in order to support those Member States who were not part of the European Union, when the Kyoto Protocol was signed and ratified, *"2% of the total quantity of allowances to be auctioned being distributed amongst Member States the greenhouse gas emissions of which were, in 2005, at least 20% below their emissions in the base year applicable to them under the Kyoto Protocol."*

For the use of the revenues from auctioning the allowances another compromise was found. *"Member States shall determine the use of revenues generated from the auctioning of allowances. At least 50% of the revenues generated from the auctioning of allowances"* shall be used for a range of climate protection measures, such as energy efficiency, renewable energy, avoiding deforestation, and Carbon Capture and Storage. Up to 50% of these measures can be offset by financing comparable measures in developing countries.

Eventually, there are regulations for protecting industries exposed to *"carbon leakage"*. According to the Emissions Trading Directive 2009, Recital 24, page 14 *"the Community should allocate 100% of allowances free of charge to sectors or sub-sectors meeting the relevant criteria"*. The European Commission shall – according to the directive – assess in detail which sectors or sub-sectors are exposed to the risk of carbon leakage, starting by end of December 2009, and then regularly reviewing the situation.

3.5   EFFORT SHARING DECISION

The *"Decision of the European Parliament and the Council on the effort of Member States to reduce their greenhouse gas emissions to meet the Community's greenhouse gas reduction commitments up to 2020"* (Effort Sharing Decision 2009) deals with emissions reduction from sectors not covered by the ETS-directive. As a starting point, the decision reiterates consensus and earlier decisions of the Community on climate change, such as the 2°C target and the need for drastic reduction of greenhouse gas emissions: *"The view of the Community, most recently expressed, in particular, by the European Council of March 2007, is that in order to meet this objective, the overall global annual mean surface temperature increase should not exceed 2°C above pre-industrial levels, which implies that global greenhouse gas emissions should be reduced to at least 50% below 1990 levels by 2050. The Community's greenhouse gas emissions covered by this Decision should continue to decrease beyond 2020 as part of the Community's efforts to contribute to this global emissions reduction goal. Developed countries, including the EU Member States, should continue to take the lead by committing to collectively reducing their emissions of greenhouse gases in the order of 30% by 2020 compared to 1990. They should do so also with a view to collectively reducing their greenhouse gas emissions by 60 to 80% by 2050 compared to 1990."* (Recital 2, page 3) The decision includes the aim to integrate maritime shipping and international aviation in efforts to reduce greenhouse gas emissions. If no international agreement is reached by the end of 2011, a Commission proposal for possible unilateral implementation shall be presented: *"Such a proposal should minimise any negative impact on the Community's competitiveness while taking into account the potential environmental benefits".*

The decision highlights existing legislation for climate protection policies and recasts in progress and then describes the intention of the European Union to *"further ensure a fair distribution between the Member States of the efforts to contribute to the implementation of the independent reduction commitment of the Community".* To achieve fairness, several rules are spelled out for effort sharing within the EU, as well as for including reductions achieved through the Clean Development Mechanism (CDM) and through Joint Implementation (JI) according to the Kyoto Protocol. Until an international agreement on climate protection has entered into force, Member States shall be allowed to use additional credits for projects in developing countries of up to 3% of their emissions not covered by the ETS-Directive and up to 1% of their verified 2005 emissions to be calculated against their reduction targets.

Based on these considerations, the decision consists of technical and political definitions and agreements about how the non-ETS sectors shall contribute to the 20% (and 30%) emissions reduction target of the EU. The decision allows Member States to carry forward, for the period from 2013 to 2019, up to 5% of their annual emission allocation (Article 3 (3), page 19). Member States in compliance with the decision may transfer up to 5% of their emission allowances to other Member States.

As a result of the different flexibility, banking and borrowing options, up to 50% of a country's reduction obligation can be transferred to other countries. Another weakness of the decision is a mechanism similar to the one in the Emissions Trading Directive 2009, which calls for a new decision process, if the overall reduction target of the EU of 30% instead of 20% becomes operational. But on the other hand, it is a logical consequence of the respective agreement in the ETS-Directive.

However, the effort sharing decision constitutes an important step forward for climate protection measures in the EU's non-ETS sectors, which account for about 60% of the Union's greenhouse gas emissions. In Annex II, it sets – for each of the EU's 27 Member States – binding national greenhouse gas reduction targets (or limits of increase of greenhouse gas emissions for those who need this in the context of a fair effort sharing) for the period from 2013 to 2020 compared to 2005 levels. The targets of Annex II add up to a reduction of about 10%[6] for the period from 2013 to 2020, which is in compliance with the overall 20% reduction target and which would have to be updated by increasing the respective reduction targets to comply with a 30% reduction commitment.

---

[6]The 10% share is mentioned by the rapporteur of the Effort Sharing Decision, the Finnish MEP Satu Hassi (Greens/EFA) in a European Parliament's Press Release (European Parliament 2008).

3.6   CCS DIRECTIVE

The "*Directive of the European Parliament and of the Council on the geological storage of carbon dioxide*" (CCS-Directive 2009) was brought forward with the intention – as the Rapporteur, Chris Davis, ALDE, UK, explained in a Press Release (European Parliament 2008) – to "*set the foundations for the development of CCS technology that will help us secure massive reductions in CO_2 emissions from power stations and industrial installations. The regulatory framework provides for the safe and permanent underground storage of CO_2, and we have provided the financial means to bring about the construction of 9 or 10 commercial CCS demonstration projects across Europe*". This quotation not only explains the main objectives of the CCS-Directive, but it also shows that the discussion about the directive was closely linked with the rapporteur's suggestion that CCS projects should be financed by public funds, in particular the revenues of the new entrants reserve of the ETS-Directive[7].

According to the agreed text of the directive (CCS-Directive 2009, Recital 4, page 3) "*Carbon dioxide capture and geological storage (CCS) is a bridging technology that will contribute to mitigating climate change [...] Its development should not lead to a reduction of efforts to support energy saving policies, renewable energies and other safe and sustainable low carbon technologies, both in research and financial terms*". The text holds that "*seven million tonnes of CO_2 could be stored by 2020 and up to 160 million tonnes by 2030, [...]. The CO_2 emissions avoided in 2030 could account for some 15% of the reductions required in the Union*".

The European Council of March 2007 had urged "*to work towards strengthening research and development and developing the necessary technical, economic and regulatory framework in order to remove existing legal barriers and to bring environmentally safe CCS to deployment with new fossil power plants, if possible by 2020*" (CCS-Directive 2009, Recital 8, page 5). In the following recitals and in the text of the directive emphasis is laid on the necessity to provide a suitable framework for financing projects, for securing environmentally safe storage and for mitigating the risks of this new technology. It is clear – and underlined in the directive – that CCS is not a mature technology, which has already proven its technical, environmental or economic viability. Therefore, the purpose of the directive is to trigger a development which should bring about the experience needed to assess whether CCS can fulfil the expectations to deliver significant greenhouse gas reductions at economically sound costs[8].

The directive is based on these assumptions and thus is intended to establish "*a legal framework for the environmentally safe geological storage of carbon dioxide (CO_2) to contribute to the fight against climate change*" (Article 1, 1). And the limitations are mentioned (Article 1, 2): "*The purpose of environmentally safe geological storage of CO_2 is permanent containment of CO_2 in such a way as to prevent and, where this is not possible, eliminate as far as possible negative effects and any risk to the environment and human health.*" It is interesting to note that for the purpose of the directive it is obviously not mandatory to prevent risks, but it is sufficient to "*eliminate as far as possible [...] risk to the environment and human health*". This certainly is an issue for lacking public acceptance of carbon storage.

---

[7]Among the negotiators there was an agreement about 500 million emissions allowances (value estimated at 10 billion Euros) to be set aside for new market entrants. The revenues from this New Entrants Reserve (NER) should be used to finance innovative projects. MEP Davis insisted and succeeded to include large scale CCS demonstration projects in the list. This was criticized for double crediting these emissions: not subject to emissions allowances according to the ETS-Directive and additionally eligible for subsidies here. This was part of a compromise, which also included eligibility of large scale Renewable Energy projects and of grid infrastructure projects.

[8]Generally, CCS is criticized for various reasons. It reduces the efficiency of power plants equipped with this technology; it increases the cost of energy production; it creates additional risks and there is little public acceptance for the infrastructure (pipelines, storage) needed for implementation. Additionally, CCS is criticized for providing a pretext for continued use of fossil energy despite environmental problems and limited availability of suitable sites for storage.

The directive underlines (page 28–29) that it is the sole responsibility and the right of the Member States to select suitable sites for storage, and also *"not to allow for any storage in parts or in the whole of their territory"*. In any case, a storage site shall only be selected (Article 4, 4), *"if no significant environmental or health risks exist"*. Chapter 3 of the directive sets up technical and administrative procedures and precautions for obtaining, holding and changing storage permits.

Chapter 4, pages 38–53, determines *"Operation, closure and post-closure obligations"*. Among other requirements, Member States have to secure regular monitoring of irregularities, migration and leakage of CO_2. Operators are obliged to regularly report the main facts of the installation. The competent authorities shall *"organise a system of routine and non-routine inspections of all storage complexes within the scope of this Directive for the purposes of checking and promoting compliance with the requirements of the Directive and of monitoring the effects on the environment and on human health"* (Article 15, 1, page 42). Based on reports and inspections, Member States shall secure that necessary measures, such as corrective actions or closing of the storage installation are taken.

In Chapter 5, *"Third-party access"* is constituted so that *"potential users are able to obtain access to transport networks and to storage sites for the purposes of geological storage of the produced and captured CO_2"*. The access shall *"be provided in a transparent and non-discriminatory manner determined by the Member State"* (page 53).

In the final provisions of the directive, the European Commission is required to provide an evaluation report by 31 March 2015, which should include proposals for potentially necessary amendments of the directive.

3.7   RENEWABLE ENERGIES DIRECTIVE

The *"Directive of the European Parliament and of the Council on the promotion of the use of energy from renewable sources amending and subsequently repealing Directives 2001/77/EC and 2003/30/EC"* (Renewables Directive 2009), in short *"Renewable Energies Directive"* is often considered to be the flagship of the Climate and Energy Package. There is a wide range of reasons for this assessment.

The Renewable Energies Directive is building on the successful Electricity Directive 2001 and the Biofuels Directive 2003. In addition, it is integrating the repeated request of the European Parliament and many stakeholders for a directive on the use of renewable energy in the heating and cooling sector.

The Renewable Energies Directive is the legal instrument for the implementation of the European Council's agreement to increase the share of renewable energy in the EU's overall energy consumption to at least 20% in 2020. The agreement included taking into account the different starting points in the 27 Member States[9], the different support mechanisms[10], the different

---

[9]The reference year for the share of Renewables in the 27 Member States (due to availability of Eurostat data) is 2005. In 2005, according to Annex I of the directive, the existing shares of renewable energy in the national energy mix of the 27 Member States varied from 0.0% in Malta, 0.9% in Luxemburg, 1.3% in the United Kingdom on the lower end to 28.5% in Finland, 34.9% in Latvia and 39.8% in Sweden on the upper end. The national overall targets for 2020 were set accordingly, with a range from 10% (Malta), 11% (Luxemburg) and 13% (Belgium, Czech Republic, Cyprus, Hungary) on the lower end to 38% (Finland), 40% (Latvia) and 49% (Sweden) on the more advanced side.

[10]The scope of support schemes in the Member States included and still includes a broad variety of different instruments. There are feed-in tariffs and feed-in premiums, quota obligations and tradable green certificates, tendering models and tax incentives, just to mention a few. They differ according to the technologies covered, the level of support, the number of years they are provided, the stability of the framework, the application in different sectors and – most importantly – they differ widely between the individual Member States. Different systems are combined in one country whereas in others they are not.

economic potentials, the different levels of market liberalization, the different levels of inter-connection to neighbouring countries and considerably differing ideas about the future energy mix. The directive had to build the bridge between Member States with well developed support instruments for at least some renewable energy technologies and applications and those with no or nearly no established support instruments so far. And it had to bridge the gap between Member States with high potentials of different renewable energy sources and/or strong ambitions to achieve high shares of Renewables domestically and those with lower ambitions and/or lower or less cost efficient potentials. Eventually, two important issues had to be addressed and the related conflicts had to be solved.

Already in the late 1990's, when the Electricity Directive 2001 was discussed and eventually passed through Council and Parliament, there had been strong lobbying for a *"harmonisation"* of support schemes[11]. At that time, this lobbying had successfully been outweighed by supporters of existing national support systems, such as the German feed-in law[12]. But, although in 2001 the conflict was settled in favour of national support systems, the advocates of *"harmonisation"* and more *"market based*[13] *instruments"* had never given up. Despite the facts clearly proving the superiority of well designed feed-in systems for the majority of analysed examples, as regards efficiency and cost effectiveness, the supporters of harmonising support for electricity from renewable sources on a European Tradable Green Certificates (TGC)/quota-system continued to strive for their interpretation of a *"market based"* support mechanism.

Supporters of *"harmonisation"* – among them some of the European Commission's staff in charge of energy policies – came close to suggesting that reality should adapt to their interpretation of economic theories: in a *"Commission Staff Working Document"* (European Commission 2008) presented to the public together with the first official draft of the Renewables Directive 2009, which – among other important findings about various support policies in the electricity sector – basically was a proof for the superiority of well designed feed-in systems, they wrote *"Economic theory has shown that under ideal conditions, quantity-based instruments and price-based instruments have the same economic efficiency"* (page 5). However, reality – including the document itself – had proven that *"price-based instruments"*, such as TGC and quota, were by far less efficient and less cost effective than *"quantity-based instruments"* (feed-in systems).

---

[11]The intention of those advocating *"harmonization"* at that time was a shift towards what they called *"market based"* instruments. For the advocates of *"harmonization"* this meant focussing on Tradable Green Certificates and/or quota systems as the theoretically more *"market based"* instruments.

[12]In Germany, an early version of a feed-in tariff system had entered into force already on 1 January 1991. The *"Stromeinspeisegesetz"* (Electricity Feed-in Act) triggered a first wave of take-off for wind power, although – due to technical, administrative and awareness related restrictions at that time – the increase was still limited. The *"Stromeinspeisegesetz"* was replaced by the *"Act on granting priority to renewable energy sources (Renewable Energy Sources Act – EEG)"* on 1 April 2000. The EEG, after some amendments, is still the principal instrument of support for electricity from renewable sources in Germany. And it is still a model for many other feed-in laws world-wide. An unofficial English translation of the version, which entered into force on 1 January 2012, can be downloaded from the webpage of the Ministry in charge of renewables in Germany, the Ministry for the Environment, Nature Conservation and Nuclear Safety (EEG 2012).

[13]For reasons, which have never been explained sufficiently, some economists insist that a quota system and tradable green certificates are *"market based"* whereas feed-in systems are not. There is no apparent logical reason, why an instrument, which defines the quantity of renewables (letting the market decide about the price), is *"marked based"*, whereas an instrument, which determines the price of renewables (letting the market decide about the quantity), is not. It seems that the fact that most of the economists favouring TGC and quota systems are supporting traditional energy utilities, is the only element which helps understanding why this is the case. On the other hand, well designed feed-in systems are easier to apply for new market entrants, such as small and medium enterprises, farmers and even private persons, who have become relevant competitors for market shares in some countries applying feed-in systems.

It was therefore not surprising that first drafts of the Renewable Energies Directive, leaked in 2007, included elements of an EU-wide certificate trading system[14]. This trading option was maintained in the official draft (Renewables Directive 2008). At the end of the deliberations in the European Parliament and the Council, when the final agreement was voted on, a broad majority once again rejected *"harmonisation"* of support instruments and insisted on national support schemes combined with newly introduced *"cooperation mechanisms"* between Member States being the key elements for successfully achieving the respective national targets and the overall target of the directive.

The second conflict to be addressed and solved was – and partly still is – the use of sustainable biofuels (i.e. those produced from sustainably grown biomass) for transport and a related binding 10% biofuels target for the EU as a whole and for each Member State, as it was developed in the draft directive. Different aspects had to be dealt with. Whereas environmental NGOs opposed the use of biofuels due to sustainability concerns particularly regarding direct and indirect land use change, others doubted the availability of sufficient resources, in particular if the use of biomass for food were considered a strong priority[15]. As a result, in the final directive the 10%-target became a target for renewable energy in the transport sector, including not only first, second and later generations of biofuels, but also renewable electricity used in electric cars and other forms of renewable energy use in the transport sector.

Others voiced their concerns about the technical aspects of biofuels use. Whereas some stakeholders argued that first generation biofuels were technically inferior and not appropriate for modern engines, some governments and – not surprisingly – car manufacturers believed that second generation biofuels would perform much better. Environmentally and technically, they are deemed to be superior, because they are using nearly all parts of the respective biomass and they can be optimized for use in highly efficient cars with very low emissions – if they are commercially available in due time. This is why the final directive has a review clause in this regard.

3.7.1   *Binding overall targets and indicative trajectories*

It has become a broad consensus not only among supporters of accelerated development and deployment of renewables that targets can effectively trigger green growth in the energy sector, in particular, if they are combined with well designed support policies. An increasing number of countries, states and provinces world-wide (96 in 2010, up from 76 in 2008) have developed related policy targets in recent years[16]. For the European Union, targets aiming at a certain share of renewables were introduced in renewable energy policies with the Electricity Directive 2001,

---

[14]From the beginning of the drafting within the Commission (as could be heard through the grapevine), these elements were designed to be applied in combination with national support mechanisms. The formal and informal debate about the negative impacts of such an EU-wide trading option was probably one of the most important aspects, if not the most important one, in the process of agreeing on and finalizing the directive.

[15]This aspect has been addressed in various studies and campaigns (e.g. Oxfam, *"Another Inconvenient Truth: How biofuel policies are deepening poverty and accelerating climate change"*, 25 June 2008; other NGOs, such as Greenpeace and WWF are still campaigning against unsustainable biofuels use, sometimes implying the impossibility of sustainable use of biofuels for significant shares of transport fuels). As a result, a consensus is developing with regard to the directive and beyond. In case of doubt, the use of land and crops for food should have the highest priority. However, the use of sustainable biofuels should not be prohibited, if sustainability can be proven by clear criteria and/or a valid certification system. There is no general agreement about how criteria or certification system should look like. Some still question whether sustainable use of biomass can ever be achieved. The sustainability criteria in the Renewables Directive 2009 are an important element of the solution, but the discussion on concrete application, e.g. for assessing indirect land use change (ILUC), is ongoing, just like the issue of extending the sustainability criteria to all forms of biomass use for energy production will have to be further discussed.

[16]More details about global policy development, policy targets and policies can be found in the *"Renewables 2011: Global Status Report"* (GSR 2011), the recent version of the report annually published by REN21, the *"Renewable Energy Policy Network for the 21st Century"*.

which set an indicative target for the European Union and obliged Member States to decide about their own indicative targets for 2010, which had to be in line with the overall target for the EU to reach a share of 12% renewable energy in the overall energy consumption, and to achieve a 22.1% share of renewables in the EU's electricity consumption. Annex I of the Electricity Directive 2001 listed reference values for each Member State (15 at the time). The directive obliged Member States to regularly report about the progress achieved and the Commission to report about the overall progress achieved towards the targets (more details in Chapter 2 of this book).

The indicative targets of the Electricity Directive 2001 triggered different levels of ambition in renewable energy support, including in the 12 Member States joining the EU after 2001. Reports from the European Commission, based on Member States reports, have shown repeatedly that not all Member States made real efforts to reach their targets[17]. The same applies to an even stronger degree[18] for the 2010 indicative biofuels targets of 5.75% for each Member States, which were set by the Biofuels Directive 2003.

Hence, the overall experience with policy targets for renewables was ambiguous, not as encouraging as could have been expected. Although target setting and related reporting did facilitate relevant progress in some Member States, particularly in the electricity sector and much less in the biofuels sector, it was evident that the absence of penalties – except public blaming – for non-compliance was a major weakness of the directives. Consequently, supporters of renewable energy[19] soon started to call for new and stronger targets beyond 2010. For the heating and cooling sector, a 25% target for 2020 emerged from the discussions and become a major aspect of several debates and decisions of the European Parliament. In 2004, the European Renewable Energy Council (EREC) launched a campaign for a 20% renewables target for 2020 (EREC 2004a), combining the overall target with sectoral assessment and related targets for electricity, heating and cooling, and for transport. This campaign was supported by EREC's member associations as well as other stakeholders and parliamentarians. In parallel to the campaign for a 20% overall target and related sectoral target, some stakeholders argued that the targets had to be legally binding so that in case of non-compliance infringement procedures could be started and result in effectives penalties, which would eventually help enforce compliance with the directive.

In the months before the meeting of European Council in March 2007, potential allies for a binding 20% minimum target and related ambitious sectoral targets were frequently approached in debates and formal and informal meetings. Despite clear support for these targets, doubts were voiced about the willingness or at least readiness of all 27 heads of state and government to agree on a binding target, and even less on a 20%-target, which some stakeholders believed to be quite ambitious. Even stronger concerns were raised about the chance to reach an agreement on ambitious sectoral targets. Some of the participants of the discussions among supporters of 2020-targets suggested to assess whether there were potential trade-offs and/or mutual exclusiveness of a binding overall target (in comparison to a new indicative target) and ambitious sectoral targets. Whereas some suggested to strive for a set of indicative targets (overall and for the three sectors) and not to risk complete failure, others believed it might be easier to focus on one binding overall target. At the end of the day, deliberations and decisions of the March 2007 Council Meeting included a binding overall target of at least 20% renewables in 2020, underpinned but mandatory overall targets for each Member State, leaving sectoral distribution and most other questions for

---

[17]Recent developments towards the 2010-targets can be found in a document published in January 2011 (European Commission 2011j). It shows which Member States were on track towards their targets of the Electricity Directive 2001 and which were not. The Renewables Directive 2009 is the immediate follow-up of this directive.

[18]According to the same evaluation (European Commission 2011j), only 5 of the 27 Member States had reached the target, 11 were far away from it with no real effort to comply, the other 11 had at least made some efforts.

[19]Numerous Members of the European Parliament and of national parliaments and in particular the Brussels based stakeholder associations called for ambitious new targets, most strongly for the electricity sector, but also for the heating and cooling sector, where neither targets nor any EU legislation existed so far.

the Member States to decide. Information about how the compromise was achieved in the Council seem to support those who had argued for one overall target and for leaving the rest to the Member States. There is some evidence that sectoral targets would not have been acceptable for some of the Heads of State and Government.

In a difficult and complex process after the Council Meeting, the European Commission suggested a list of differentiated targets for each Member State, which was included later as Annex I of the Renewables Directive 2009, which was eventually agreed on and thus became legally binding. The targets were defined as shares of gross final energy consumption. The Commission's proposal was based on a range of criteria, such as technical and economic potential of each Member State, existing shares of renewables in each Member State in 2005, which was the latest available Eurostat data at that time, and the economic potential of each Member State.

In addition to the binding overall targets, Annex I defines *"indicative trajectories"*[20], which had been introduced – as *"interim targets"* – into the debate, in order to secure a smooth and credible development towards the 2020 targets. There were many who believed the trajectory should also be mandatory so that efficient infringement procedures could be started earlier and with stronger impact. With all the different interests of the different Member States, an agreement on binding interim targets and even less so on direct penalties for remaining below the trajectory was not in the focus of the various debates before the directive eventually passed. However, the Commission repeatedly underlined their view that they would certainly be willing and legitimized by the directive to start infringement, as soon as a Member State falls below the indicative trajectory. It remains to be seen, if Member States will proactively strive to comply with the trajectory or even be more ambitious and how many of them will actually reach the respective shares and how many will exceed the trajectory and the binding national 2020-target.

3.7.2   National support schemes and cooperation mechanisms

One of the most important issues raised with the Renewables Directive 2009 is the integrity of national support schemes. As mentioned above, there have long since been attempts to remove or at least undermine and phase out national support schemes and replace them by a European-wide *"harmonised"* mechanism. This debate was of utmost importance for the successful agreement on the directive and on potentially ambitious implementation. European renewable energy associations, environmental NGOs, Members of the European Parliament and of national parliaments all over Europe, and also the governments of Member States, which had developed and implemented efficient and cost effective support mechanisms for electricity from renewable sources, in particular Germany, Spain, Denmark and some others clearly voiced their opposition against direct or indirect *"harmonisation"* of support schemes[21]. It is worthwhile mentioning that a turning point

---

[20]The Renewables Directive 2009 foresees in *"B. Indicative trajectory"* a gradually increasing share of renewables in the Member States' gross final energy consumption, with one third of the target to be achieved in the last two years (2019 and 2020). As an average of the two-year period of 2011 and 2012, 20% of the 2020-target should have been achieved, whereas for 2013/2014 the increase should be 30% of the target, 45% for 2015/2016 and 65% for 2017/2018.

[21]The debate about advantages of national support schemes is covered in more detail in Chapter 6 of this book. Therefore, it should be sufficient here to highlight that although in the beginning of the discussion there was an open confrontation between supporters of Quota Systems (favouring harmonization) and supporters of Feed-in Systems (opposing harmonization), it soon became clear that – for the purpose of this debate – this was not the main conflict. The underlying question was, whether or not Member States can control the impact of their support policies and thus clearly follow the subsidiarity principle of the EU treaty, or whether they do not have the means to achieve their binding national targets due to interfering and potentially conflicting *"harmonized"* instruments. It is interesting to know that it was a non-paper in a Council working group on energy, jointly produced and introduced in the debate by the UK (mainly applying quota and TGC, with some feed-in support for small installations), Poland (applying quota and TGC) and Germany (applying a feed-in system), which eventually paved the way for the agreement to secure national support systems and introduce cooperation mechanisms for those who wish to cooperate with their neighbours.

Table 3.2.   National renewables shares in 2005 and 2020-targets.

Member State	RE-share in 2005 (%)	RE-target for 2020 (%)
Belgium	2.2	13
Bulgaria	9.4	16
The Czech Republic	6.1	13
Denmark	17.0	30
Germany	5.8	18
Estonia	18.0	25
Ireland	3.1	16
Greece	6.9	18
Spain	8.7	20
France	10.3	23
Italy	5.2	17
Cyprus	2.9	13
Latvia	32.6	40
Lithuania	15.0	23
Luxembourg	0.9	11
Hungary	4.3	13
Malta	0.0	10
The Netherlands	2.4	14
Austria	23.3	34
Poland	7.2	15
Portugal	20.5	31
Romania	17.8	24
Slovenia	16.0	25
The Slovak Republic	6.7	14
Finland	28.5	38
Sweden	39.8	49
United Kingdom	1.3	15

in the debate was reached, when – during a meeting of the International Feed-in Cooperation (IFIC)[22] in spring 2008 – the German government, supported by Spain, Denmark and others explained that they are not willing to accept harmonisation of support systems in the Renewable Energies Directive, because this would definitely undermine the viability of Germany's successful feed-in system and similar systems in other countries. Instead they suggested introducing flexibility mechanisms for those Member States which insisted on being unable to achieve the targets of the future directive domestically. In parallel, similar deliberations had taken place in the European Parliament, where the rapporteur for the directive, Luxembourg's Green MEP Claude Turmes, in close cooperation with the shadow rapporteurs from all other major parliamentary groups, including Social Democrats, Liberals and Conservatives, was working hard to craft an alliance in support of national support systems and the subsidiarity principle. Based on this discussion, the *"cooperation mechanisms"* were designed and fine-tuned. As a result, the finally agreed directive clearly states the need for Member States remaining in control of their support schemes.

---

[22] http://www.feed-in-cooperation.org/wDefault_7/index.php is the webpage of International Feed-In Cooperation (IFIC). Officially, only Germany, Spain and Slovenia were members of the IFIC. Greece joined in 2012. But all other EU Member States and even some countries beyond the EU are regularly participating in the meetings. The IFIC played an important role in the debate about *"harmonization"* of support systems in the Renewables Directive 2009. IFIC meetings became an important forum for exchange of views of nearly all Member States about how to design efficient support mechanisms.

For understanding how precisely the agreement on the directive was achieved and eventually formulated, it is useful to read Recital 25, which – after stating that most Member States apply some sort of support mechanisms – clearly underlines: *"For the proper functioning of national support schemes it is vital that Member States can control the effect and costs of their national support schemes according to their different potentials. One important means to achieve the aim of this Directive is to guarantee the proper functioning of national support schemes, as under Directive 2001/77/EC, in order to maintain investor confidence and allow Member States to design effective national measures for target compliance. This Directive aims at facilitating cross-border support of energy from renewable sources without affecting national support schemes. It introduces optional cooperation mechanisms between Member States which allow them to agree on the extent to which one Member State supports the energy production in another and on the extent to which the energy production from renewable sources should count towards the national overall target of one or the other. In order to ensure the effectiveness of both measures of target compliance, i.e. national support schemes and cooperation mechanisms, it is essential that Member States are able to determine if and to what extent their national support schemes apply to energy from renewable sources produced in other Member States and to agree on this by applying the cooperation mechanisms provided for in this Directive"* (Renewables Directive 2009).

These ideas are clearly specified in the normative articles of the directive. Article 3 *("Mandatory national overall targets and measures for the use of energy from renewable sources")*, No. 3 explicitly mentions support schemes and cooperation mechanisms as instruments for reaching the mandatory targets. Articles 6–11 are dealing with the *"cooperation mechanisms"*, which Member States may apply on a voluntary basis to assist each other in reaching the targets of the directive.

Article 6 defines *"Statistical transfers between Member States"*, which is the basic cooperation mechanism, and which seems to be the easiest to apply. *"Member States may agree on and may make arrangements for the statistical transfer of a specified amount of energy from renewable sources from one Member State to another Member State"*. Such transfers will be added to the amount of energy calculated towards the target of the beneficiary and deducted from the amount produced in the other Member State. Transfers have to be formally notified to the European Commission and they are only permitted if they do not affect the target achievement of the Member State making the transfer.

Article 7 and Article 8 deal with *"Joint projects between Member States"*. This seems to be the mechanism which is most often referred to in public discussions. *"Two or more Member States may cooperate on all types of joint projects relating to the production of electricity, heating or cooling from renewable energy sources. That cooperation may involve private operators"* (Article 7, No. 1). Like statistical transfers, the projects have to be notified to the Commission, including information about the respective shares of energy to be calculated towards the target of which of the involved Member States.

Article 9 and Article 10 deal with *"Joint projects between Member States and third countries"*. This element had been widely criticized during the debate in Parliament, because it opens a backdoor for cooperation in difficult contexts, which may not always be fully traceable and verifiable. Two or more Member States can cooperate with one or more third countries, and like in Articles 7 and 8, they may include private operators. However, projects with third countries can only be counted towards the targets of the Member States, if certain conditions are met. Some of the conditions are meant to be quite strong. For example, electricity, which is to be counted towards the importing country's target, has to be physically transported into and consumed in the European Union (not necessarily in the country, which is counting it towards the national target). Exceptionally, imports can also be counted in the EU, if physical interconnectors with sufficient capacity are not available before 2022. Of course, third country projects also have to be notified to the Commission with all details and attributions.

Probably the most interesting and innovative cooperation mechanism was introduced in Article 11 *("Joint support schemes")*. This article is rather a suggestion than a regulation, opening more possibilities to flexibly apply the directive and comply with its targets. Accordingly, *"two or more Member States may decide, on a voluntary basis, to join or partly coordinate their national support scheme"*. Apart from demanding detailed notification about the cooperation and potential distribution rules for the energy produced under such a cooperation structure, this article leaves the Member States a wide range of freedom in designing cooperation and/or coordination. It may be due to this openness or felt lack of concreteness that so far no two[23] (or more) Member States have actually started such cooperation, or the differences between national support schemes are deemed to be considerable barriers. This article bears a lot of potential for sound and reasonable convergence of support schemes, because it explicitly stimulates discussion and cooperation among Member States (not only, but preferably) with similar or even (nearly) identical support systems. Therefore, it could be a trigger for efficient cooperation – preferably, but not only – among neighbouring countries with mutually open electricity markets and similar or identical levels of ambition for renewable energy development and deployment.

3.7.3   *National Renewable Energy Actions Plans (NREAPs)*

It has been shown that in various Member States the Biofuels Directive 2003 and the Electricity Directive 2001 have not been implemented with sufficient ambition, resulting in missing[24] the EU's respective 2010-targets of 22% in the electricity sector and 5.75% of biofuels. This is not an unexpected outcome. And it is a major reason why stakeholders and Members of the European Parliament insisted on introducing an effective instrument in the directive which would enable the European Commission to review Member States' sectoral targets, trajectories and policies. The *"National Renewable Energy Action Plans" (NREAPs)* are the instrument which was designed to meet these expectations. Details can be found in Article 4 of the Renewables Directive. Accordingly, the European Commission had to provide – by 30 June 2009 – a template[25] for the NREAP which was to be completely answered by each Member State. The template guides the Member States through a comprehensive list of questions regarding existing installations of renewable energy technologies in the three sectors, targets and projections for the three sectors and for different technologies, description and analysis of existing support mechanisms, existing and newly introduced policies, remaining cost and non-cost barriers and many other useful information on which sound policy development should be based. Member States had to submit their NREAPs by the 30 June 2010, the latest[26].

In addition, each Member State had to submit a *"Forecast Document"*[27] indicating basic elements of the NREAP and particularly answering the question if and to what extent they were planning to use the *cooperation mechanisms* for target compliance or if they foresee exceeding

---

[23] In 2011, Sweden and Norway started a joint green certificate scheme, which – although first ideas about it are much older – practically applies some of the points of Article 11 of the Renewables Directive 2009. However, as Norway is not a Member of the European Union (but – like all EU-Member-States – a Member of the European Energy Community), it remains true that no two EU-Member States have implemented this Article so far.

[24] Although only preliminary figures for 2010 were available, it seemed likely that the EU would not meet the targets. The European Commission assumed that the share of renewables in the electricity sector would be around 19.4% and around 5% in the transport sector (European Commission 2011i).

[25] The template was made available to the Member States in time and in their national languages. It is available at European Commission's website *Transparency Platform*: http://eur-lex.europa.eu/LexUriServ/ LexUriServ.do?uri=CELEX:32009D0548: EN:NOT (viewed 30 April 2013).

[26] In reality, the last NREAP was only submitted in January 2011, but experts believe that this was not a bad track record. All NREAPs in the original language and in English can be found on the European Commission's *"Transparency Platform"* and the following link: http://ec.europa. eu/energy/renewables/transparency_platform/action_plan_en.htm

[27] The Documents are available on the European Commission's Transparency Platform Webpage: http://ec.europa. eu/energy/renewables/transparency_platform/forecast_documents_en.htm

their targets so that they can offer a certain amount of energy for statistical transfer or other cooperation agreements.

If Member States fail to reach their respective indicative trajectories, the directive obliges them to resubmit their NREAP, indicating clearly how they plan to fill the gap in the following years. The European Commission shall evaluate the action plans and – if necessary – recommend changes.

The process of drafting the NREAPs was of great importance for meaningful implementation of the Renewables Directive 2009. Based on this assessment, several national renewable energy associations had designed a project which became known by its acronym *REPAP2020*[28]. The project was developed and conducted in close cooperation with the European Commission's civil servants in charge of renewable energy, and the implementation of the diverse related directives and potential infringement procedures.

With the very much appreciated and accepted support for drafting the NREAPs and with the evaluations elaborated and presented in the project, significant data and expertise was made available for a sound assessment of the NREAPs, when and to a large extent even before they were eventually submitted. The project also analysed – immediately after their submission – the NREAPs for their comprehensiveness and for their levels of ambition. In March 2011, the project published a brochure providing a European *"Industry Roadmap"* and a first evaluation of the NREAPs, showing that – according to the sample of submitted NREAPs a share of more than 20% (20.7%) Renewables in 2020 was likely. According to industry calculations, a considerably higher share of 24.4% seems achievable (EREC 2011a).

More details about policies, targets and projections developed in the 27 NREAPs and an assessment of their impact will be analysed in Chapter 4 of this book.

3.7.4   *Guarantees of Origin (GOs)*

Apart from renewable energy production triggered by support instruments on different levels, a voluntary market for *green electricity* has emerged in recent years. Based on the willingness and on the intention of an increasing number of citizens to decide for themselves about the quality of the energy consumed and to opt for renewables, including the readiness to pay higher prices and to choose a related tariff for their electricity consumption, a specific market segment is emerging, with only clean energy being sold. In this context *"clean energy"* is used to describe the electricity offered by green energy providers. In most cases this is 100% renewable power, with different shares of new installations included. In some of the tariffs, a certain share of fossil power produced in CHP (Combine Heating and Power) plants is included. *"Clean energy"* in this sense should not be confused with the use of some advertising campaigns of incumbent utilities, including *"clean coal"* or *nuclear* in an energy mix they call *"clean"* or *"low carbon"*. And of course, the green quality of the electricity provided has to be proven without reasonable doubt. This is why certificate systems were created and applied in various countries, providing proof of *"greenness"* of the electricity in question.

An important aspect is the reliability of such certificates, the question whether or not they offer reasonable security against fraud of any kind, and particularly against double counting,

---

[28] *"Renewable Energy Policy Action Paving the Way towards 2020"* (REPAP2020) was a project supported by the European Commission's Intelligent Energy Europe (IEE) Agency and coordinated by EREC and EUFORES. The project established cooperation between European and national renewables associations and scientists in order to accompany the NREAP development and offer support for those Member States who were willing to accept. In addition, the consortium developed national and European *"Industry Roadmaps"* for comparison of policies and ambition levels and suggested changes to the NREAPs where this was deemed appropriate. Close cooperation with MEPs, coordinated by EUFORES, the European Parliamentarians' Forum for Renewable Energy Sources, was another element of the project. Eventually the NREAPs were evaluated, assessed and policy recommendations were developed for further improving them. Details of the project and the studies and assessments produced can be found on the project webpage: http://www.repap2020.eu/

which is a real challenge for credibility and acceptance of such a certification system. It is evident that consumers should only once be charged potentially higher prices for green electricity. A reliable and valid certification system – with the purpose of proving greenness of the product and protection against fraud – is not easy to design and even more difficult to enforce. This is a major reason, why quite some expertise was needed to develop the system of *"Guarantees of origin" (GOs)* as it is now incorporated in the Renewables Directive 2009[29].

The directive includes a clear and comprehensive framework about what should be the purpose of GOs and what should be avoided (Renewables Directive 2009, Recital 52): *"A guarantee of origin can be transferred, independently of the energy to which it relates, from one holder to another. However, with a view to ensuring that a unit of electricity from renewable energy sources is disclosed to a customer only once, double counting and double disclosure of guarantees of origin should be avoided"*. According to the directive GOs have to be electronic documents which are mutually recognized and accepted by all Member States. This is elaborated in detail in Article 15 (*"Guarantees of origin of electricity, heating and cooling produced from renewable energy source"*): GOs are standardized to 1 MWh and have to be issued upon request. It has to be ensured that only one GO is issued for each unit of energy produced and that a GO is used only once. Member States or their designated competent bodies have to supervise the use and the cancellation after use. In their biannual reports about progress achieved towards the targets of the directive, Member States have to include information about reliability of the GOs and of their fraud resistance.

3.7.5 Removing barriers

The range of the Renewables Directive is much broader than commonly known. Apart from targets and cooperation mechanisms, it includes detailed regulations about transparent procedures and removal of administrative and other remaining barriers against renewable energies.

The directive requires streamlining administrative procedures and defining clear timetables so that the process is transparent and non-discriminatory. For small installations, complex authorisation procedures should be replaced by simple notification wherever possible. Article 13 outlines in detail which deliberations and regulations are necessary for a smooth development of renewables. Authorities are required in detail to provide transparency and sufficient information, clear timelines, proportionate procedures taking into account the particularities of different technologies. European labels and standards shall be applied wherever possible.

The need for information to be available for all interested actors, consumers and suppliers, architects and installers is highlighted. Article 14 lists information and training needs. Member States are obliged to provide information, certification systems and guidance for relevant actors, such as consumers and local and regional authorities.

Smooth and reasonable grid integration of renewables is another important aspect. The directive obliges Member States to provide clear and favourable rules for grid access of renewables: *"Priority access and guaranteed access for electricity from renewable energy sources are important for integrating renewable energy sources into the internal market in electricity ..."* and *"the objectives of this Directive require a sustained increase in the transmission and distribution of electricity produced from renewable energy sources without affecting the reliability or safety of the grid system. To this end, Member States should take appropriate measures in order to allow a higher penetration of electricity from renewable energy sources, inter alia by taking into account the specificities of variable resources and resources which are not yet storable. To the*

---

[29]It should be mentioned that some stakeholders believed (and some still do) that these Guarantees of Origin (i.e. certificates) should also be used as instruments proving target compliance according to the Renewable Directive 2009. However, as the market for certificates is a separate market from the electricity market, this would have a number of negative effects on the stability of support systems and as a consequence on public acceptance for renewable energy deployment. However, as the directive allows the certificates only for proof of greenness ("disclosure"), this aspect is mentioned here without need for further elaboration.

extent required by the objectives set out in this Directive, the connection of new renewable energy installations should be allowed as soon as possible" (Recital 60).

The directive prescribes that costs for grid connection shall be transparent and non-discriminatory and taking into due account the benefits of renewables for the electricity and gas grid (Recital 62). Article 16 deals in detail with important aspects of *"Access to and operation of the grids"*. It establishes the obligation for Member States to *"take the appropriate steps to develop transmission and distribution grid infrastructure, intelligent networks, storage facilities and the electricity system, in order to allow the secure operation of the electricity system as it accommodates the further development of electricity production from renewable energy sources, including interconnection between Member States and between Member States and third countries. Member States shall also take appropriate steps to accelerate authorisation procedures for grid infrastructure and to coordinate approval of grid infrastructure with administrative and planning procedures"* (Article 16, No. 1). In the following text, article 16 outlines in detail the need for priority or guaranteed access to the grid and for priority dispatch of energy from renewable sources. Grid operators have to make transparent their rules for bearing and sharing costs. Technical rules and costs for connection to the electricity and gas grids have to be published. Member States are obliged to address the problem of frequent curtailment of renewable energy sources for reasons of grid stability. *"Member States shall ensure that appropriate grid and market-related operational measures are taken in order to minimise the curtailment of electricity produced from renewable energy sources. If significant measures are taken to curtail the renewable energy sources in order to guarantee the security of the national electricity system and security of energy supply, Members States shall ensure that the responsible system operators report to the competent regulatory authority on those measures and indicate which corrective measures they intend to take in order to prevent inappropriate curtailments"* (Article 13, No. 2c). This regulation will certainly become increasingly important – and so will monitoring of enforcement – in the coming years, when the shares of Renewables – particularly variable resources such as wind and solar – in Europe's electricity mix will significantly increase. Without proactively tackling this challenge – the need for enabling the grids to accommodate increasing shares of Renewables by grid-enforcement and enhancement, deployment and development of storage capacity and installation of smart grids – the shift towards a renewable energy based system will be more difficult and more costly than necessary.

In addition to tackling barriers, the directive includes clear language and important regulations about increasing energy efficiency, in order to more easily achieve the renewables targets and to contribute to climate protection. Combining energy efficiency and renewables is definitely the way forward. And due to the fact that legislation on efficiency – on European and on national levels – is much less developed than legislation and progress on renewables, it is all the more urgent that the Renewables Directive is addressing this context.

3.7.6   *The heating and cooling sector*

One of the very important tasks, which the Renewables Directive had to facilitate, was the development and integration of European policies for heating and cooling from renewable sources. Following decisions of the European Parliament, a directive – similar to the Electricity Directive 2001 and the Biofuels Directive 2003 – should be developed for the heating and cooling sector. Council and European Parliament explicitly demanded to include this objective in a comprehensive Renewables Directive.

Following the agreement of the Council and the European Parliament to set binding overall targets for renewable energy and to leave it to the Member States to decide how these overall targets should be distributed between the three sectors (as long as the overall target is met as well as the 10% minimum target for renewables in transport), the Renewables Directive does not specify quantitative targets for Renewables shares in the heating and cooling sector. But it includes a number of regulations and policies fostering Renewables in this sector. It requires that *"Member States shall introduce in their building regulations and codes appropriate measures in order to increase the share of all kinds of energy from renewable sources in the building sector. [...] By 31 December 2014, Member States shall, in their building regulations and codes or by other means with equivalent effect, where appropriate, require the use of minimum levels of energy from renewable sources in new buildings and in existing buildings that are subject to major renovation [...]"* (Article 14, No. 4). Public buildings shall serve as examples and therefore fulfil these requirements already two years earlier, beginning in January 2012 (Article 14, No. 6).

3.7.7 Renewable energy in the transport sector

In addition to the binding overall targets for each Member State, the directive also constitutes – for each Member State and for the Union as a whole – a binding 10% minimum target for 2020 and renewables in the transport sector. As described earlier, the European Commission had proposed to set the target to a 10% share of biofuels (Renewable Energies Directive 2008). Following environmental and technical concerns an agreement was reached that other forms of renewable energy use in the transport sector – particularly electromobility – should be taken into account and become part of the directive and should be accountable for target compliance. The 10% target is now described as a share of renewable energy used in the transport sector, with certain exceptions (some experts call it inconsistencies) for aviation and marine transport.

When calculating the share of renewables in transport, a specific calculation rule is established (Renewables Directive 2009, Article 3, No. 4). Whereas only fuel and electricity consumed in road and rail transport is taken as the denominator, the numerator includes all energy consumed in all forms of transport. This may be acceptable for reasons of incentivising the use of Renewables outside road and rail transport, but if biofuels or electric engines become more frequent in planes and ships, this will result in a figure which is much higher than the actual share of renewables in the respective country's transport energy mix.

And there is another promotional aspect in the directive regarding electricity consumed for transport. Building on the difficulty to exactly assess the share of Renewable sources contributing to electricity consumed in cars and offering a calculatory increase for some Member States, the directive allows choosing between two calculatory paths. *"Member States may choose to use either the average share of electricity from renewable energy sources in the Community or the share of electricity from renewable energy sources in their own country [...]"*. Following this sentence, Member States with a lower share of renewables in the electricity sector than the EU's average may apply the higher EU average for the sake of achieving – mathematically – a higher share of renewables in transport.

It seems to be evident that for the period up to 2020 the lion's share of renewables in the transport sector will be biofuels, predominantly first generation – despite the possibility to count eletromobility's share for target compliance. This is why the aspects of sustainability on the one side and the availability of second generation biofuels on the other side were and still are so important in the discussion about renewable energy in the transport sector.

3.7.8 Sustainable biofuels and other biomass

Following concerns about unsustainable biofuels production and consumption and alleged (and sometimes real) competition between the use of biomass for food or for energy production, there is a broad agreement that only sustainable biofuels should be accountable for the targets of the directive. To achieve this aim, the directive includes three specific articles about sustainable biofuels. When the directive eventually passed the Council and the European Parliament, 24 recitals (more than for any other aspect of the directive), three comprehensive and detailed articles and a 21-page-annex were included, all concerning sustainability of biofuels.

The recitals clearly outline the rationale behind the directive's biofuels regulations. Biofuels should be sustainable, and only sustainable biofuels can count for target compliance. In order to avoid unsustainable biofuels being used for other forms of energy production, the sustainability criteria have to apply to all bioliquids (Recitals 65, 66, 67). The directive's Article 17 *"Sustainability criteria for biofuels and bioliquids"* underlines that only bioliquids meeting the sustainability criteria may be counted towards the national targets of the directive. As a basic overall approach the directive allows only for liquids with at least 35% of greenhouse gas savings compared to fossil fuels to be included in target reaching. As of 2017 this figure increases to 50% and – for installations with production starting after 1 January 2017 – to 60% in 2018.

To be considered sustainable and thus be accountable, bioliquids cannot stem from particularly biodiverse land and they must not incentivize destruction of such land (Recital 69). Furthermore, the effects of transforming land with high carbon stocks into arable land and related direct and indirect effects have to be taken into consideration (Recitals 70–73). Therefore Article 17 No. 3, 4 and 5 explicitly lists where bioliquids taken into account for the purpose of the directive must not have been grown. The enumeration covers woodland which has not been affected by human influence so far, most protected areas, a major part of highly biodiverse grassland, and land with high carbon stocks.

Recital 74 requires taking into account negative effects of bioliquids produced outside the EU, where it might be difficult to prove that sustainability criteria are properly respected. Eventually, the Commission is required to analyze the effects of biomass use for other energy applications (Recital 75). The European Union should strive for extending criteria and credible certification to other parts of the world (Recital 79). For a sound assessment of positive and negative effects of bioliquids, the directive elaborates in detail on the calculation of greenhouse gas emissions (Recital 80, Article 19 and Annex V: *"Rules for calculating the greenhouse gas impact of biofuels, bioliquids and their fossil fuel comparators"*). The Annex contains average or typical values for greenhouse gas savings to be applied for various feedstocks and also for fossil fuels. A compromise is proposed between detailed fact based assessment and regional average in order to avoid unnecessary administrative burden. Article 19 *("Calculation of the greenhouse gas impact of biofuels and bioliquids")* is the legal basis for the application of the calculations detailed in Annex V.

The directive calls for incentives to prioritize more diversification and thus more sustainable biofuels. In order to achieve the target of sustainable and reasonable biofuels use in an efficient way, the directive implies some changes in the fuel quality directive, which was debated and decided in parallel with the Renewables Directive, allowing for higher blending rates of biofuels to conventional petrol or diesel. And of course, regular reporting is required and established. Renewables Directive 2009, No. 6 and 7 set up detailed requirements for the Commission's reports on these issues.

The present directive does not include sustainability criteria for energetic use of biomass other than biofuels. But the problem is addressed by asking the European Commision to present, by end of 2009, a report analysing this area and presenting a proposal for further dealing with the related problems and challenges.

Article 18 defines the requirements for Member States on and economic actors for setting up verification schemes and mass balance analysis. And it outlines the need for negotiating and agreeing with third countries on the application of the sustainability criteria. By the end of 2012, the Commission shall report on the effectiveness of the system and on potential extension to other criteria in relation to air, soil or water protection (Article 18, No. 9).

3.7.9   *Review in 2014*

Like most European directives the Renewables Directive is a compromise between different interests and different assessments. It has, nevertheless, become a very well developed and promising tool. It is not unusual that directive includes a number of reports which Member States and the European Commission have to present at defined dates. And it is not unusual that the directive includes a review clause, foreseeing that at a certain point in time a debate is envisaged about potential amendments to the directive.

The Commission and the Member States have to submit a number of reports about achievements and potential solutions for questions which were left open in the directive – be it for a lack of evidence or for a lack of consensus.

Member States have to submit progress reports every second year, the first one was due on 31 December 2011; the last one will have to be prepared by 2021. Following the logic of the directive and particularly of the template for the National Renewable Energy Action Plans, Article 20[30] describes in great detail what kind of information the report has to include, be it related to target achievement, enacted policies, remaining barriers, availability of sustainable biomass or to potentially necessary corrective actions.

Building on the Member States' reports, the European Commission has to present a report every second year describing and analyzing the overall progress achieved in the Community along the indicative trajectory and towards the 2020-targets. The Commission's report has to specifically analyze the biofuels sector with a view to the various issues related to sustainability in the context of bioliquids and biomass. For 31 December 2014, the directive foresees an extended report to be submitted by the Commission, which is commonly understood to be a review clause for major elements of the directive. The requirement for the Commission's report includes – among several other aspects – a review of the greenhouse gas thresholds as outlined in the directive, assessment of the availability of sustainable biofuels with a focus on second generation and – which is probably the most challenging part for the political discussion – an assessment of the effectiveness of the cooperation mechanisms and the national support schemes.

Based on the report the Commission shall, if deemed appropriate, submit proposals for adaptation of greenhouse gas saving requirements and adjustments to the cooperation mechanisms. With regard to the controversial debate about the specific benefits of national support schemes, it is not surprising that there is an explicit clarification trying to prevent re-opening the harmonisation debate only a few years after the directive's entering into force: *"Such proposals shall neither affect the 20% target nor Member States' control over national support schemes and cooperation measures"* (Article 23, No. 8). Early in 2013, when this book was written and finalized, it had become evident how important this precautionary text of the directive is[31]. The debate was ongoing. And it is very likely that it will continue – at least – until a political consensus about a post-2020 climate and energy framework has been achieved.

3.8   THE IMPORTANCE OF THE CLIMATE AND ENERGY PACKAGE

The Climate and Energy Package has rightly been called a landmark of European energy policies. Despite some weaknesses and critical aspects it is a fair assessment of the set of legislative acts taken in the package. However – and this is in line with debates in Europe and beyond about the effectiveness of specific policies for sustainable energy development – when it comes to detailed assessment of more or less successful implementation and related effectiveness for shaping the transformation of the energy system towards a future oriented sustainable path, there is not only

---

[30]Following the broad acceptance of the Template for the NREAPs the European Commission provided a detailed template for the Member States' reports due at the end of 2011. It can be downloaded in all EU languages from the Commission's Transparency Platform: http://ec.europa.eu/energy/renewables/transparency_platform/template_progress_report_en.htm. At the same webpage all reports submitted by the Member States can be found in the original language and in English.

[31]It may be political irony that it was the newly appointed Energy Commissioner, the German Christian Democrat Günther Oettinger, who reopened the debate about in 2010, when he presented his first strategic paper as the European Commissioner for Energy, only a few months after taking office. In a Communication (European Commission 2010) he addressed various topics to be thoroughly considered and elaborated in the energy arena until 2020, including support schemes for renewable energy. This strategy paper and other aspects of the present Commission's policies with regard to renewables will be analyzed in Chapter 11 of this book.

a range of strong opinions, but there is a set of facts to be observed – showing that some parts of the package seem to be well developed and successfully being implemented and others are not. In the following chapter, a first assessment will be tried about the implementation of the package, showing that the Renewables Directive is probably the most efficient and effective part, delivering smoothly on the targets and triggering discussions about new and more ambitious targets for the time beyond 2020, whereas energy efficiency policies and the Emissions Trading Directive and even more the CCS Directive will still have to prove the positive impact they might have on providing secure and sustainable energy and on progress towards effective climate protection.

CHAPTER 4

From agreement via legislation to implementation – will the climate and energy package deliver until 2020?

Jan Geiss

4.1 INTRODUCTION

The European Union (EU) has established a set of targets for renewable energy, energy efficiency and reduction of greenhouse gas emissions for the year 2020. Whilst the transposition and implementation are work in progress one can assume that the political frameworks on EU and national levels are in the first place able to deliver those 2020 targets. However, a wide spread lack of ambition leads to the conclusion that all three targets are still under threat.

Renewables are on track in the year 2013, but the planned growth rates are going up towards 2020 and one has to fear that most Member States lack the willingness to really implement those targets.

The 20% energy efficiency target for 2020 will not be reached from the perspective of 2013. The new energy efficiency directive has changed the situation, but more ambition and legislation is necessary to reach the 20% in 2020.

The greenhouse gas emission reduction of 20% for 2020 lies in the success or failure of the European emission trading system (EU-ETS). The first phases of implementation have shown that systemic errors lead to a complete failure of the system so far. Carbon prices remain low so that ETS does not deliver incentives strong enough to create emission reductions.

All in all, the success of the climate and energy package lies in the hand of political institutions. If decision makers see the opportunities in the implementation of the 2020 targets, success is possible. However, if Member States, their leaders and voters continue to see those measures as a burden, the three 2020 targets will most probably be missed.

4.2 THE CLIMATE AND ENERGY PACKAGE

The European Union's Climate Package has created the grounds for a new coherent and forward looking European policy on sustainable energy solutions. The 2020 targets have established an interim step towards a fully sustainable energy system.

The Climate Package intends to reach by 2020:

- a minimum level of renewable energy sources in the final energy consumption of at least 20%,
- a 20% target of energy efficiency,
- a 20% target for the reduction of greenhouse gas emissions.

This chapter describes and evaluates the latest status of implementation of the three policy areas, gives an assessment about future developments and an answer to the core question: will the three sectors reach the defined targets as it stands now or do further improvements need to be established?

One basic problem about forecasting remains: based on the agreements, legislation and implementation of the climate package parts – can a forecast seriously be done about real fulfilment of the goals or can a judgement be made of a potential failure of the climate package? The text will elaborate on this and give an assessment on the main reason to failure: a Europe-wide lack of political willingness and shortcomings in the design of implementation tools, which intend to make the climate package happen on the ground and on European energy and technology markets.

In addition to the general political and scientific debate on the climate package, this chapter is based on the work of EUFORES[1] and on comprehensive evaluations which have been done by the following high level European projects of EUFORES and its partners:

– *REPAP 2020 – Renewable Energy Policy Action Paving the Way towards 2020*[2] – a joint effort of a broad spectrum of European and national stakeholders, scientific institutes and legal advisors – among them EUFORES – the European Forum for Renewable Energy Sources and project co-ordinator EREC – the European Renewable Energy Council. Its mission was to facilitate the implementation of the Renewables Directive and to support the development and the roll-out of the national renewable energy action plans.
– *Keep-on-Track!*[3] analyses early market feedback and estimates, if EU Member States are on track concerning the target trajectory of the RES-Directive
– EEW1 and EEW2 – *Energy Efficiency Watch*[4] – a project lead by EUFORES with the mission to assess EU and national energy efficiency policies and action plans as well as gathering expertise from market actors on the successes and failures of energy efficiency policies all over Europe.

4.3   STATUS AND PROSPECTS: RENEWABLE ENERGY SOURCES

4.3.1   *Legislative background: The RES-Directive*

With the Renewables Directive (Renewables Directive 2009), as described in Chapter 3 of this book, the European Union has created a stable legislative framework for the promotion of renewable energy sources[5]. The directive defines the target for the EU to get at least 20% of the EU's final energy consumption from renewable energy sources in 2020 (as compared to an 8.5% level in 2005). Binding national targets of RES shares have been established for each Member State's gross final energy consumption together with a share of 10% renewables in the transport sector.

As a key instrument to achieve the targets, the directive requests the use of National Renewable Energy Action Plans (NREAPs, defined in Article 4 of the directive), which had to be submitted by the EU Member States to the European Commission by 30 June 2010. EU Member States are expected to define in those NREAPs how they plan to reach their national goals in the electricity, heating and cooling and transport sectors.

All Member States had sent their respective NREAPs to the European Commission early in 2011. They are published on the European Commission Transparency Platform.[6] The European Commission did an encompassing evaluation of the NREAPs. As a first result, several infringement procedures were initiated against Member States who had delivered insufficient action plans in the eyes of the Commission. Thus, the Commission intends to push national legislators to come up with improved NREAPs.

The key criteria for judging a NREAP as insufficient were (EUFORES 2011):

– Incomplete implementation of the directive,
– Significant deviation from plan or trajectory,
– Valid complaints from any EU citizens regarding incorrect implementation.

---

[1] EUFORES is a European cross-party and non-profit network of Members of Parliament from the European Parliament and the EU27 national parliaments with the mission to promote renewable energy and energy efficiency. http://www.eufores.org
[2] http://www.repap2020.eu
[3] http://www.keepontrack.eu
[4] http://www.energy-efficiency-watch.org
[5] "Renewable Energy – What do we want to achieve", European Commission, downloaded 13.2.2013, http://ec.europa.eu/energy/renewables/index_en.htm
[6] "Renewable Energy – Transparency Platform", European Commission, downloaded 13.2.2013, http://ec.europa.eu/energy/renewables/transparency_platform/transparency_platform_en.htm

4.3.2 Status: The National Renewable Energy Action Plans – NREAPs

The analysis shows that based on the submitted NREAPs[7], nine countries will meet their national targets by their own resources and instruments: Belgium, Cyprus, Estonia, Finland, Ireland, Latvia, Portugal, Romania and the UK. 16 countries have announced to overshoot the target and to produce a surplus beyond their minimum target: Austria, Bulgaria, Czech Republic, Denmark, France, Germany, Greece, Hungary, Latvia, Malta, the Netherlands, Poland, Slovenia, Slovakia, Spain and Sweden. Only two countries fear not to reach their targets: Italy and Luxembourg plan to make use of the cooperation mechanisms, which are established in the directive and described in Chapter 3 of this book. In total, Member States plan to reach 20.7% in 2020 and thereby to slightly go beyond the 2020 20% target (EUFORES 2011).

Table 4.1 shows – in addition to the individual national targets as defined by the RES-Directive – the target as defined by each country in their respective NREAP and forecasts done by European renewables industry within the REPAP2020 project to illustrate the national potentials.

The evaluation of the national renewable energy action plans – as done by Fraunhofer ISI[8] and EEG Vienna[9] within the REPAP2020 project – shows that the NREAPs are of very different quality and completeness. "*While some countries submitted complete and comprehensive roadmaps to 2020 outlining the status quo and the remaining issues to be tackled, other countries focus on the policies in place which do not always match reality. In some cases, the NREAP template was even insufficiently filled in. The scientific evaluation was conducted in five categories: administrative procedures and spatial planning, infrastructure development and electricity network operations and support schemes for heating and cooling, electricity and transport. As a conclusion, it was found that the most room for improvement was detected in the field of administration and spatial planning, and adequate support measures in the heating and cooling sector. From the evaluation of the NREAPs as well as from the feedback received from national stakeholders under the REPAP2020 project, the general picture is that many well-known bottlenecks in the further deployment of RES also prevail in the NREAPs*" (Eufores 2011).

In addition, conclusions and recommendations have been drawn and outlined in the REPAP2020 Policy Recommendation Report (BBH 2010), which stresses the shortcomings of the NREAPs:

- "*Proposed trajectories and planned measures are not aligned;*
- *Member State lack of ambition and/or commitment through action;*
- *Lack of administrative knowledge and trust in RES as a viable, reliable set of energy solutions in each sector;*
- *Administrative barriers such as burdensome and complex permitting procedures;*
- *Discrimination under various tax regimes;*
- *Grid constraints in the form of poor operations, lack of guaranteed and priority access and dispatch, and/or lengthy grid connection times;*
- *Low support levels in some of the Member States, especially in the RES-Heating & Cooling sector;*
- *Emphasis on maintaining status quo growth or, at best, incremental change rather than concrete measures targeted at significant, large gains, e.g. as few as nine Member States plan RES building obligations or comparable measures;*
- *Support that does exist, does not target the full-spectrum of different RE technologies in each of the sectors; Lack of public information and knowledge;*
- *Access to data is weak and unreliable;*

---

[7]"Renewable Energy – Action Plans & Forecasts", European Commission, downloaded 13.2.2013, http://ec.europa.eu/energy/renewables/action_plan_en.htm
[8]http://www.isi.fraunhofer.de/isi-en
[9]http://www.eeg.tuwien.ac.at/

Table 4.1. Individual national targets (Source: EREC 2011a).

Country	National target as defined by the RES-Directive [%]	Target as defined by the NREAP [%]	Renewable energy share in final energy consumption forecast by the RES industry (based on demand assumptions from the NREAPs) [%]
Austria	34	24.2	46.4
Belgium	13	13	14.5
Bulgaria	16	18.8	20.8
Cyprus	13	13	14.5
Czech Republic	13	13.5	13.7
Denmark	30	30.5	30.5
Germany	18	19.6	26.7
Estonia	25	25	25
Greece	18	20.2	25.2
Spain	20	22.7	28.3
Finland	38	38	42.3
France	23	23.26	23.6
Hungary	13	14.7	18.3
Ireland	16	16	16
Italy	17	16.2	19.1
Lithuania	23	24.2	31.7
Luxembourg	11	8.8	10.4
Latvia	40	40	46.4
Malta	10	10.2	16.6
Netherlands	14	14.5	16.8
Poland	15	15.5	18.4
Portugal	31	31	35.3
Romania	24	24	24
Slovenia	25	25.2	34.1
Slovakia	14	15.3	26
Sweden	49	50.2	57.1
United Kingdom	15	15	17
Total EU	20	20.7	24.4

- 'Stop-and-go' nature of many national policies are detrimental to the industry/ investors on all levels; and
- Some economies have been hit especially hard by the financial crisis which has resulted in banks decreasing financing/lending for projects, including RES."

It appears that several countries have taken the obligation to provide a complete and comprehensive RES roadmap up to 2020 seriously, indicating what is in place and what needs to be done. Others drew a nice picture on the implemented measures to stimulate an enhanced RES deployment which does not in all cases match with reality. Very few countries provided a minimalistic and incomplete report (FHI/EEG 2011).

The overall summary of the evaluations for the countries analysed by the REPAP project is presented in Table 4.2. It can be seen that substantial optimisation potentials exist for all five categories. The strongest deficits exist in the field of administrative procedures and spatial planning followed by the category support measures for RES heating and cooling. The highest optimisation potentials exist in these two areas. But even the section on support measures in the electricity sector receives only a neutral evaluation on average showing room for improvement in many EU Member States.

Table 4.2. Overall assessment of the NREAPs (Source FHI/EEG 2011).

Topic Country	Administrative procedures and spatial planning	Infrastructure development and electricity network operations	RES electricity support measures	RES heating and cooling support measures	RES transport support measures
Austria	☺	😐	😐	😐	☺
Belgium	😐	😐	😐	☹	😐
Bulgaria	☹	😐	😐	☹	☺
Cyprus	☹	😐	😐	😐	☹
Czech Republic	☹	😐	😐	😐	😐
Denmark	☺	☺	😐	😐	☺
Finland	☹	😐	😐	😐	😐
France	☹	☹	☺	😐	☺
Germany	☺	😐	☺	😐	😐
Greece	😐	☹	☺	☹	☹
Ireland	😐	☹	😐	😐	☺
Italy	☹	☹	☺	😐	😐
Latvia	☹	☹	😐	😐	😐
Lithuania	☹	☹	😐	😐	☹
Malta	☹	😐	☹	☹	😐
Portugal	☹	☺	😐	☹	😐
Romania	☹	😐	😐	☹	☹
Slovenia	☹	😐	☺	😐	😐
Spain	😐	😐	😐	😐	😐
Sweden	☹	☺	😐	☺	☹
United Kingdom	😐	😐	😐	☹	😐

As an additional source of judgement on the future developments the European Commission has published its assessment on the cooperation mechanisms as foreseen in the Renewables Directive in Article 4(3). This forecast shows how Member States intend to fulfil their interim and final targets by domestic means or by the use of cooperation mechanisms. This is already a good indication on how serious Member States see their own technical and political means to reach the targets.

A good overview is provided by the European Commission's summary of the Member States' NREAPs[10]: *"In accordance with Article 4(3) of Directive 2009/28/EC on the promotion of the use of energy from renewable energy sources, all Member States have submitted documents giving their forecast of the expected use they will make of the cooperation mechanisms contained in the Directive. The cooperation mechanisms include 'statistical transfers' where Member State governments can agree to exchange statistically a given quantity renewable energy produced. Another mechanism is the 'joint project', where a specific new plant is identified and the output of the plant shared statistically between Member States. Joint projects concerning electricity production can also be established with third countries if a number of conditions are met, most importantly if the electricity is physically consumed in the EU. The intention behind the Directive's creation of these instruments is to allow Member States to achieve their targets in a cost effective manner, developing renewable energy sources wherever it is most efficient to do so. [...]*

---

[10]Summary of the Member State forecast documents, European Commission, downloaded 13.2.2013, http://ec.europa.eu/energy/renewables/transparency_platform/doc/dir_2009_0028_article_4_3_forecast_by_ms_symmary.pdf

Key findings from the reports are:

- *At least ten Member States expect to have a surplus in 2020 compared to their binding target for the share of renewable energy in their final energy consumption. This surplus could be available to transfer to another Member State. The quantity is estimated at around 5.5 Mtoe, or around 2% of the total renewables needed in 2020.*
- *Spain and Germany forecast the largest surpluses in absolute terms, with 2.7 Mtoe and 1.4 Mtoe respectively.*
- *Five Member States expect to have a deficit in 2020 compared to their binding target for the share of renewable energy in their final energy consumption. These Member States thus require transfers from another Member State or third country, through the use of the Directive's cooperation mechanisms. The quantity amounts to around 2 Mtoe (<1% of the total renewable energy needed in 2020).*
- *Italy forecasts the largest deficit in absolute terms, of 1.2 Mtoe.*
- *The net result of Member States' forecasts for 2020 renewable energy consumption is that the EU should exceed its 20% target by over 0.3 percentage points.*
- *The comparatively small quantity of energy expected to be subject to the cooperation mechanisms reflects most Member States' ability to develop domestic resources cost effectively and their desire to reap the economic social and environmental benefits of developing renewable energy sources nationally. However it remains the case that the cooperation mechanisms created by the Directive are available should Member States wish to make further use of them and achieve their targets even more cost effectively.*
- *A total of 13 Member States also expect to exceed the interim targets that result from the trajectory contained in the Directive and thus have a surplus in the years before 2020. Three Member States anticipate a deficit during this period. Thus Member States may also use the cooperation mechanisms to meet their trajectory in the years before 2020. (It is worth recalling that the Directive requires Member States to plan to meet or exceed their trajectory).*
- *Many Member States point out that these trajectories and targets require strong, new national energy efficiency and infrastructure measures."*

Another source of judgement on the latest state of the renewables situation towards 2020 is a communication by the European Commission as published in summer 2011 (European Commission 2011i). Its key conclusions and recommendations were:

"The limited and fragmented growth of Europe's renewable energy industry in the decade to 2008 resulted partly from the limited EU regulatory framework. Recognising that renewable energy will form the heart of any future low carbon energy sector, the EU introduced a comprehensive and robust supportive legislative framework. The challenge is now to move from policy design to implementation at national level, with concrete action on the ground. The implementation of the Directive and the presentation of plans are encouraging signs of progress that need to be sustained. In the current context of macro-economic fragility and fiscal consolidation, it is important to recognise the financing for renewable energy as growth-enhancing expenditure that will provide greater return in the future. It is equally important to ensure the quality of the expenditure, applying the most efficient and cost effective financing instruments. As with energy infrastructure, there is a need for European action, to speed up the efficient delivery of renewable energy production and its integration into the single European market. At national level, any revision of financing instruments should be pursued in a way that avoids creating investor uncertainty and takes into account other Member States' policies to ensure an approach coherent with the creation of a genuine European market. The Commission will actively support national cooperation on financing renewables, based on the new framework for Member State cooperation contained in the Renewable Energy Directive and promote the integration of renewable energy into the European market. At European level, EU funds should be directed to ensure cost effective renewable energy development and providing technical assistance while ensuring the most effective means of lowering the cost

of capital investments in the sector, including in collaboration with the EIB [Auth.: European Investment Bank] and provision of technical assistance. The Commission therefore invites Member States to

- *implement the National Renewable Energy Action Plans;*
- *streamline infrastructure planning regimes while respecting existing EU environmental legislation and strive to conform to best practice;*
- *make faster progress in developing the electricity grid to balance higher shares of renewable energy;*
- *develop cooperation mechanisms and start integrating renewable energy into the European market;*
- *ensure that any reforms of existing national support schemes will guarantee the stability for investors, avoiding retroactive changes.*

To support such efforts the Commission will continue to work in partnership with Member States on the implementation of the Directive, to review and improve the effectiveness of EU funding for renewable energy projects and facilitate the convergence of national support schemes in order to ensure the best conditions for the development of renewable energy in Europe."

The European Commission has thereby already well analysed many shortcomings and risks to failure concerning the future implementation of the Renewables Directive and the development of renewables market shares in the EU.

4.3.3 *Policy recommendations*

Based on the analysis and the conclusions drawn out of the work within the REPAP2020 project and the ongoing scientific work and market players' feedback in the "Keep-on-Track"! project, a couple of policy recommendations are given for the electricity, heating and cooling as well as the transport sector. They shall help to improve the quality, structure and ambition of national renewables policies and to push the implementation of the RES-Directive to the necessary level with the aim to reach at least the 20% renewables target for 2020:

- Ensuring constant growth rates of renewables along the RES-Directive trajectory in all EU Member States until 2020;
- Stable RES support schemes to create investors security and to support young technologies until they are able to compete independently;
- Exploiting a wide range of technologies to create a diversified base for the future energy mix and to keep up the run to the best solutions for a certain period without stopping innovation processes too early;
- Reliable statistics as a fair basis for political decisions and for market comparison;
- Efficient administrative procedures and spatial planning to avoid non-monetary barriers to the market development of renewables;
- Effective electricity infrastructure development and operation to guarantee access of renewables to the grids and to lay the ground for a both regional and European grid and a decentralised grid suitable to central and decentralised renewables solutions in a heterogeneous mix of sources;

As a conclusion, the report states:

"The manifold benefits of the deployment of renewable energy are undeniable. The increasing generation of energy from renewable sources will not only help the EU to meet its greenhouse gas emission targets but also reinforce the transformation into a sustainable economy. Meanwhile, it has positive impacts on the security of energy supply and the overcoming of the dependence on energy imports.

The binding renewable energy target for 2020 serves as an accelerator for the promotion of renewable energy and the National Renewable Energy Action Plans are crucial instruments helping Member States to reach their targets. Nevertheless, the over-arching conclusion of the in-depth analysis of the NREAPs can only be that – despite the outlook that the overall target will be met – old bottlenecks persist and more vigilance and guidance is needed" (EUFORES 2011).

With Directive 2009/28/EC the European Union has created a progressive, clear and committed legislative framework to achieve a minimum of 20% of renewables in the overall energy use by 2020. The sum of the binding national targets will add up to the minimum overall target of 20%. The directive thus created a *"community of positive fate"* between the Member States (BBH 2011).

It is of utmost sensitivity that the Commission will sit at the steering wheel when it comes to assist and partly guide the process of realization. Meanwhile, the Commission can rely on a multitude of concrete information in order to facilitate its work even though more strategic work is needed. Policies for increasing the deployment of renewables without a strategic approach for an energy system change encompassing the whole demand and supply chain will not be robust and not create a forward looking sustainable energy policy.

The study of the various policies of the Member States shows a lot of positive examples in various States – right down to the communal and regional level, planning and authorization. No country is perfect, even strong frontrunners do have weak points, be it in view of imbalanced support mechanisms, be it in view of robustness of all sectors involved, e.g. electricity, heating/ cooling and transport. The seriousness of all Member States to deliver will be a crucial point to discover during the coming months and years. The EU needs more "positive story" telling.

The whole endeavour to change the face of Europe's energy system towards a sustainable renewable driven and energy efficient world needs a much stronger reality commitment of all Member States and the EU level to decrease energy consumption and to increase energy efficiency as fast as possible.

4.3.4   *Status: implementation of the RES-Directive*

As a recent update on the situation, the European Commission has published a communication in summer 2012 giving feedback and a judgement on the latest state of the renewables situation in the EU (European Commission 2012c). It repeated there the request to show strong ambition in implementing the NREAPs on the national and local level. It reconfirmed the need to develop a clear perspective for infrastructure developments and to ensure investment security in order to create a stable development of renewables investments and markets in the EU.

However, it has to be criticized that the Commission has almost completely left out the heating and cooling sector based on renewables. Has it been a huge progress with the RES-Directive to add heating and cooling based on renewables to the electricity and the transport sectors, it now seems that institutions in Brussels tend to weaken the role of heating and cooling via renewables even though heating and cooling represents about 50% of EU's energy consumption. It has to be judged as a rather dangerous development for the 2020 renewables target that focus is only put on the electricity sector. A failure of the target achievement is under this development almost inevitable.

On the status of the interim target achievement, the situation seems to be positive in the first place. All experts agree that EU Member States are in general on track, if it comes to the first two years of the implementation of the trajectory and the interim targets. This is for sure a good start. However, this is just a snapshot and doesn't indicate the trends for the future. In fact, the trajectory curve starts very low and without bigger ambition, while it gets steeper and steeper in the later years towards 2020. The real ambition will lie in those last years, where growth rates have to be higher and renewable energy installations have to catch up quite fast. Therefore, the picture cannot be that positive anymore at the moment because most Member States don't show bigger ambition to promote renewables to the necessary level and fears are justified that many of them will have problems to reach the growth rates necessary to really reach the 2020 national goals.

4.3.5   *Conclusions: prospects for renewable energy*

The financial and economic crisis hits renewables support hard. The fact that many Member States see investments into renewables more as a burden than as an investment into future technologies, jobs and energy independence etc. gives reason to fear that RES support will rather remain weak or even be diminished in the coming years – several retroactive measures in several Member States show that recently (Spain, Portugal, Greece, Czech Republic, etc.).

One has to recommend therefore that – despite the economic difficulties – Member States shall rather start implementing and building renewables at larger scales earlier on in order to have the investments and security of business ready to reach the 2020 goals. A broad variety of options which do not rely on state budgets are available to tackle the still existing market barriers and to stabilise and increase investments into renewables. The positive effects would go beyond the mere development into renewables: it lays the ground for economic growth, new green technology markets, job creation, positive health and environment effects and in the end less cost for Europe citizens (details in ECN 2011 and in European Commission 2011i).[11]

Member States need to define mid- and long-term strategies for the development of our energy systems and the way we use energy. It is clear that the definition of the transition is crucial in order to make investments effective – into technologies, companies and infrastructure.

Beyond the general market crisis, specific sector difficulties can also be seen which have to be addressed as well. The recent crisis of the photovoltaic markets for example shows very clearly the lack of direction in Europe's policies: Member States and European Institutions fail to deliver a clear industrial strategy which could lay the ground for a sustainable growth of renewables industries in Europe which are necessary to make the renewables story a success story not only in the EU. New market designs will be needed in order to integrate renewables into the existing or future markets in a way, that it automatically rewards services like energy production, secondary services like grid balancing, storage and back-up systems by market mechanisms. However, today such a market design is far from being implemented and EU even lacks ideas how to make such markets happen. More research and discussion will be needed to define such a new comprehensive market design.

Also, a wake-up call for the heating and cooling sector needs to be made: renewables are not catching up speed there and in many discussions, this key sector is completely forgotten and effective measures seem to be rare to push the development into the right direction with measurable progress in market shares and installations.

To sum up: the legislative background for a strong renewables development has been laid with the renewables directive and with most national action plans. However, the European Institutions together with EU Member States will have to show more efforts and dedication to make sure that the development of the renewables markets and the achievement of the interim and final targets of the Renewables Directive will be secured. Huge efforts will be necessary to keep Europe on track with its targets.

4.4   STATUS AND PROSPECTS: ENERGY EFFICIENCY

Energy efficiency can make a significant contribution to three key policy priorities of the European Union: climate protection, energy security and the technology leadership of EU industry. The second target of the climate package is the energy efficiency target of 20% energy savings in 2020 compared against projections for 2020. The main intention of the target and its supporting policies is the decoupling of the economic development of the EU from the energy use by making behaviours, systems, buildings, transport, processes and products more and more energy efficient: do more with less.

---

[11] http://ec.europa.eu/energy/renewables/targets_en.htm is the comprehensive website of the European Commission with information about the 2020 renewables targets.

4.4.1   Legislative background – energy efficiency policies

The European Union has a manifold history of legislation dealing with energy efficiency. Several sectors have been tackled to raise the level of energy efficiency in industry, households and the service sector. The following EU directives had been established before summer 2012:[12]

- Directive on energy end-use efficiency and energy services – "Energy Services Directive", "ESD", Directive 2006/32/EC;
- Energy Performance of Buildings Directive, "EPBD", Directive 2002/91/EC on the energy performance of buildings, Recast of the EPBD in 2010 Directive 2010/31/EU;
- Cogeneration Directive, Directive 2004/8/EC on the promotion of cogeneration based on a useful heat demand in the internal energy market;
- Energy efficiency of products:
  - Product energy consumption: Information and labelling (from July 2011);
  - Tyre labelling;
  - Ecodesign for energy-using appliances;
  - Ecodesign requirements for fluorescent lamps, for high intensity discharge lamps, and for their ballasts;
  - Household appliances: energy consumption labelling (until 2011);
  - Energy efficiency of office equipment: The Energy Star Programme (EU – US);
  - Hot-water boilers.

Surely, some progress could be seen through those directives, e.g. the efficiency classification of household appliances, new rules for more energy efficient buildings etc. And yet, the climate package target of 20% in 2020 was far out of reach. Despite many efforts of stakeholders and institutions in the EU, progress in the Member States was in general too weak. Estimates – before the new Energy Efficiency Directive came into place – had shown that at the current rate, the EU was expected to meet not even half of its 20% by 2020. Only in 2012, the European Institutions finalised the new directive which improves the picture (Energy Efficiency Directive 2012).

4.4.2   Status of the efficiency policies – gaps and remaining policy requirements

But before one can judge on the future development of energy efficiency in the EU, the recent situation and success or failures should be understood. A Europe wide project and survey has gathered experts' feedback on the situation: The Energy-Efficiency-Watch project (EEW) has analysed the perception all over Europe on progress or shortcomings of energy efficiency policies and their actual implementation.

In 2006, the European Union adopted the directive on energy end-use efficiency and energy services (ESD 2006), which was so far one of the most important energy efficiency related European frameworks. The ESD was the one policy measure to harness the benefits of energy efficiency for the European society. The purpose of this directive was to make the end use of energy more economic and efficient. According to the ESD, Member States had to adopt and achieve an indicative energy saving target of 9% by 2016. It set obligations on national authorities as regards energy savings and energy efficient procurement, and measures to promote energy efficiency and energy services. The ESD required Member States to submit three National Energy Efficiency Action Plans (NEEAPs), scheduled for 2007, 2011 and 2014. In their NEEAPs, Member States had to show how they intended to reach the 9% indicative energy savings target by 2016 and describe the planned energy efficiency measures. Furthermore, NEEAPs had to show how Member States comply with the requirements of the ESD on the exemplary role of the public sector and the provision of information and advice to final consumers. Additionally, in the first NEEAPs (2007), Member States had to set intermediate targets, and the second (2011) and third NEEAP (2014) had to show results of the energy efficiency improvements and savings made.

---

[12] From "Summaries of EU Legislation: Energy Efficiency", European Commission, downloaded 13.2.2013, http://europa.eu/legislation_summaries/energy/energy_efficiency/index_en.htm

The main findings and conclusions of the Energy-Efficiency-Watch project based on the experts' feedback show the main issues which prevent a breakthrough in energy efficiency improvements. They can serve as a basis for better policies and implementation activities in the future. The report (EEW 2012) states as key requests:

– Stronger political will and more ambition whilst understanding the benefits of energy efficiency;
– New and ambitious frameworks of energy efficiency policies establishing new European leadership;
– Improved governance and policy management: multi-level governance with clearer responsibilities and more staff in executing ministries and agencies;
– More ambition in the transport sector where progress is the weakest;
– Intensified building renovation and establishment of more effective financing tools;
– Understanding for the broad spectrum of barriers other than money and the complexity of energy efficiency and its Europe-wide deployment.

4.4.3   *The new Energy Efficiency Directive*

In 2012, the European Institutions created the new EED (Energy Efficiency Directive 2012). It replaces the Energy Services Directive and the Cogeneration Directive and combines energy efficiency aspects all along the energy chain – from the first steps of energy conversion, through transport and distribution down to the level of energy use in the different sectors of society.

One key political element of the directive is the fact that – again – Member States did not want a binding target for energy efficiency for 2020, but they agreed to binding measures as an alternative. European Commission agreed to this under the precondition that if in 2014 not enough efforts could be shown by the EU Member States, the indicative target for 2020 could become binding and thereby legally enforceable.

4.4.3.1   *Key elements of the Energy Efficiency Directive*
The key characteristics of the EED are the non-binding target, binding measures, its sectoral approach, its general measures, new measuring and reporting tools. When the directive will take force on 1 January 2014, the Member States will have to renovate each year 3% of the total floor area of heated and/or cooled buildings owned by their central government. Furthermore, each Member State will have to set up an energy efficiency obligation scheme ensuring that energy distributors and/or retail energy sales companies will achieve by the end of 2020 a cumulative end-use energy savings target of 1.5% of the annual energy sales to final consumers. This obligation scheme still leaves the option to the Member States for flexibility measures as well as equivalent alternative measures. The text of the directive contains also provisions on energy audits and energy management systems, metering, billing information and promotion of efficiency in heating and cooling, energy transformation, transmission and distribution, and energy services. Also, an obligation on each EU Member State was introduced to draw up a roadmap to make the entire buildings sector more energy efficient by 2050 (commercial, public and private households included).

The EED is expected to create hundreds of thousands of jobs in Europe in the short term and in the long run contribute significantly for reaching the EU's climate objectives and decrease the EU's dependence on fossil fuels.

The key elements of the EED are follows:

– National targets and binding measures;
– National Energy Efficiency Action Plans – to be delivered every 3 years from 30 April 2014;
– Efficiency in energy use:
  • Building renovation;
  • Exemplary role public sector;
  • Renovation in public sector buildings;

- Purchasing of the public sector;
- Obligation schemes for energy utilities;
- Energy audits and management schemes;
- Metering;
- Billing information;
- Consumer information;
- Penalties.
– Efficiency in energy supply:
  - Heating and cooling;
  - Energy transformation, transmission and distribution.
– Horizontal provisions:
  - Qualification, accreditation and certification schemes;
  - Information and training;
  - Energy services;
  - National energy efficiency funds.
– Review and monitoring of implementation.

4.4.3.2   Remaining gaps

The Energy Efficiency Directive is clearly a big step forward towards the 20% efficiency target for 2020. However, as it has been shown, it will not deliver the full 20%, but rather 15% if implemented and followed correctly.[13] On top of that, proposed measures on boilers and on cars and vans have been promised which will take the EU to 17% efficiency in 2020.

One has to be aware that legislative frameworks are in the first place texts and paper and that there might be huge inconsistencies with the realities out there in the Member States – as it has been widely the case in the past. Therefore, both on the legislative level as on the practical level of implementing a comprehensive system of measures, more efforts are needed in order to reach the 2020 efficiency target in real life and measurable. Discussions on how to fill the gap with additional legislation and measures have begun in the European institutions.

Member States did not agree on a binding target – they have instead agreed on an indicative target of 20% energy savings and on legally binding measures. This is expected to result in a reduced 15% total energy savings by 2020, well short of the 20% goal that Member States had previously agreed on in principle in 2007. To make up for the shortfall, the 15% will be complemented by fuel efficiency regulation for cars and new standards for products such as boilers, which will be added to the Ecodesign Directive. This brings EU savings to 17%. The rest of the percentage will be calculated as follows:

– In April 2013, Member States are expected to present their national efficiency programmes and calculate what target they are to achieve. The European Commission will then evaluate them.
– If the Commission analysis of the national energy saving plans shows that the EU is not on track to meet the 20% energy savings target, it must add to the directive more binding measures to fill the gap.
– If Member States do not apply the additional measures and are still not on track to meet the target, Commission will then propose binding targets.
– The savings will be calculated as of 2014 and there will be a review of the directive in 2016.

4.4.4   Conclusions: prospects of energy efficiency

The EU wants to achieve a 20% energy efficiency target in 2020. In the past, first steps had been done which lead the EU to a level of about 10% in 2020. The new Energy Efficiency Directive

---

[13]"Parliament gives final green light to energy efficiency directive", Euractiv, 2012, downloaded 13.2.2013, http://www.euractiv.com/energy-efficiency/european-parliament-gives-final-news-514732

has pushed the limit to a level of about 17%. The remaining gap still needs to be filled in order to reach the 20%. Strong efforts need to be taken soon on EU, national, regional and local level if the EU wants to succeed also in the field of energy efficiency.

So far, one has to judge that the EU will fail in reaching the 20% target if no further measures are taken. Especially on the European level, time is running out having in mind that the legislative processes come to a halt for at least a year due to the elections to the European Parliament and the renewal of the European Commission[14]. But the gap has not only to be filled on the European legislative level: Even more, real measurable implementation has to take place pushed by strong commitment of all stakeholders and by a constant learning process about the shortcomings of the efficiency policies so far.

4.5 STATUS AND PROSPECTS: GREENHOUSE GAS REDUCTION

The third target of the EU climate package is the greenhouse gas reduction target to lower emissions by 20% until 2020. The main tool to reach this goal is the European emission trading system (EU-ETS).[15] The present ETS-Directive is described in some detail in Chapter 3 of this book.

4.5.1 *The logic and functioning of the EU Emissions Trading System*

The main logic of the EU-ETS is to fix the amount of total annual emissions of certain emission intensive sectors in the EU and to let markets decide about the actual price of emission allowances. This cap would be reduced each year in order to enforce the reduction of total emissions in line with international UN climate agreements (Kyoto Protocol, etc.). Under this cap, market actors would choose either to buy EU emission allowances to be able to emit, or to invest into emission reductions. The EU-ETS is an international system for trading greenhouse gas emission allowances and it covers today more than 11,000 power stations and industrial plants in 31 countries, as well as airlines. If the system functions as foreseen by its creators, 21% of emission reductions should be reached in the sectors covered by EU-ETS in the year 2020 compared to the starting point in 2005. The EU-ETS covers about half of EU's total emissions (all sectors).

The establishment of the emission trading system took place in several phases which added each time new and more concrete elements to make sure the EU-ETS works.[16]

4.5.1.1 *Phase 1: 2005–2007 – "Testing phase"*
The first phase was the so-called "testing phase", where the EU-ETS and all related services and institutions were established and tested. European Commission didn't really expect specific greenhouse gas emissions in that period.

In phase one the EU-ETS covered only emissions from power generators and energy-intensive industrial sectors. Almost all allowances were given to businesses free of charge. The penalty for non-compliance was € 40 per tonne. Phase one was successful in creating a price for carbon in the earlier stages, free trade in emission allowances across the EU and the necessary infrastructure for monitoring, reporting and verifying actual emissions. Due to the lack of reliable emissions statistics, caps in phase one were set on the basis of best guesses. In practice, the total allocation of EU-ETS allowances exceeded demand heavily and in 2007 the price of allowances fell to zero.

---

[14]Every five years, citizens of the European Union elect the European Parliament and its more than 700 Members. In addition, a new European Commission is called in. Having in mind the election campaigning beforehand and the time to get all systems back up running, the political system needs about a year to reboot its legislative capacities and to start new legislative processes which usually take 6–18 months.
[15]"The EU Emissions Trading System (EU-ETS)", European Commission, 2013, http://ec.europa.eu/clima/policies/ets/index_en.htm
[16]"EU-ETS 2005–2012", European Commission, 2013, http://ec.europa.eu/clima/policies/ets/pre2013/index_en.htm

Therefore, despite the trial character, the functioning of the EU-ETS was already then criticised after not having delivered market prices high enough to create incentives to reduce greenhouse gas emissions (IMF 2007, Chapter 1 *"Global Prospects and Policy Issues"*).

4.5.1.2   Phase 2: 2008–2012 – "Serious business phase"

In phase two, several additional elements of EU-ETS were established or improved. More gases were included, some non-EU countries joined and the penalty for non-compliance was increased to € 100 per tonne. Phase two coincided with the first commitment phase of the Kyoto Protocol. The aviation sector was also included from 2012, but its implementation put on hold until 2013 in order to give international aviation the chance to adapt to the new regulation and to find an international agreement on climate targets in aviation.

But again, the EU-ETS did not deliver prices high enough to create incentives for emission reductions. Due to the economic crisis, emission reductions happened as such just because of the reduced economic activities, and not because of a well-functioning EU-ETS. Critics have even stressed that the EU-ETS has created a huge amount of cost for the allowances, but the impact was rather zero.[17]

4.5.1.3   Phase 3: 2013–2020 – "Improve-to-deliver phase?"

Without further action, the structural surplus in allowances of the EU-ETS and the failure to create an adequate allowance price would persist also for phase three. Additional emission reductions would not be reached in an adequate amount.[18]

Therefore, European Commission has suggested several improvements in order to make EU-ETS a successful tool leading to dedicated and additional emission reductions and justifying the mere existence of the system.[19]

One of the most contentious issues in the climate change talks is a proposal that from 2013, *"there should be full auctioning of permits for energy generators and a progressive shift to full auctioning for all other industries"*.[20] Another suggestion is to block certain amounts of allowances by postponing their auctioning for a longer period in order to reduce the amount of total allowances to be issued early in phase three (900 million allowances from the years 2013–2015 until 2019–2020). In addition, a growing share of allowances will not be given for free to market actors, but distributed via an auctioning mechanism, creating incentives to well calculate emission allowance needs, their price and their future development among market actors and those companies under the EU-ETS. In 2013, 40% were auctioned and this share will increase each year. In addition, more greenhouse gases have been integrated, more sectors involved and the link with international emission allowance markets has been strengthened.[21] In addition, total emission allowances will not be fixed by national allocation plans – as it was the case in phase one and two, but a single European cap is established.

Maybe the most effective measure would be to strengthen the emission savings target from 20% to 30% in 2020 putting much stronger signals on the markets and giving room for real additional emission savings which are not caused by the economic crisis or the effects which stem from the renewables and the efficiency targets.

---

[17]"Europe's $287bn carbon 'waste': UBS report", Sid Maher, The Australian, 2011, http://www.theaustralian.com.au/national-affairs/europes-287bn-carbon-waste-ubs-report/story-fn59niix-1226203068972 – Downloaded 1 May 2013.

[18]"Emissions trading", Wikipedia, Downloaded 13.2.2013, http://en.wikipedia.org/wiki/Emissions_trading

[19]"Structural reform of the European carbon market", European Commission, 2013, http://ec.europa.eu/clima/policies/ets/reform/index_en.htm

[20]"Hanging in the balance: The EU's climate agenda", Euractiv, 2012, Downloaded 13.2.2013, http://www.euractiv.com/climate-change/hanging-balance-eu-climate-agenda/article-177751

[21]"International carbon market", European Commission, 2012, downloaded 13.2.2013, http://ec.europa.eu/clima/policies/ets/linking/index_en.htm

Overview on the discussed improvements:[22]

- Increasing the EU's greenhouse gas emissions reduction target for 2020 from 20% to 30% below 1990 levels;
- Retiring a certain number of phase three allowances permanently;
- Revising the 1.74% annual reduction in the number of allowances to make it steeper;
- Bringing more sectors into the EU-ETS;
- Limiting access to international credits;
- Introducing discretionary price management mechanisms such as a price management reserve.

If all those measures were implemented, EU-ETS would have a good chance to deliver real measurable and additional results justifying its existence, its complex institutions and structures and the cost related to it. Only then, this target and its main tool EU-ETS would make sense in addition to the renewables and the efficiency targets.

So far, with the third period of the implementation of the EU-ETS, European Commission had suggested a backloading of allowances from the beginning of the third phase towards the end in order to artificially reduce the traded allowances from 2013 on and thereby to establish a more effective carbon price on a higher level. Unfortunately, parts of the political groups in the European Parliament as well as many EU Member States are heavily against any correction of the system fearing unbearable burden for European industries in times of crisis. This gives reason to doubt that the EU-ETS will be healed in time and that it can create additional greenhouse gas emissions.

4.5.2 *Conclusions: prospects of greenhouse gas reductions*

Thus, the EU-ETS has not proven yet its full effectiveness. Still, some factors have to be adjusted in order to reach additional and measurable emission reductions.

The carbon prices had just been too low because the availability of total allowances was too high compared to the reduced real emissions in the EU. The economic crisis had reduced the total emissions, but the ETS total cap had not been adapted or designed in a way that it creates a dynamic adaptation in order to reach adequate price levels for allowances and to create incentives for market actors under ETS to invest into emission reductions.

Today, the existing EU-ETS – even in the third phase – does not look promising yet when it comes to reaching the 2020 target of reducing emissions to 20%. EU-ETS might have cost $287 billion through to 2011 and had an "*almost zero impact*" on the volume of overall emissions in the European Union and the money could have resulted in more than a 40% reduction in emissions if it had been used in a targeted way, e.g., to upgrade power plants – a disastrous judgment on the impact of EU-ETS.[23]

At least, after the disappointing results of the first two phases of the ETS – "*a disastrous track record*" (Gilbertson and Reyes 2009) – EU institutions seem to have understood that certain measures have to be taken to deliver the expected impact of the EU-ETS and the greenhouse gas emission target.

Several remaining issues continue to question the EU-ETS – even if EU institutions improve certain factors:

- The system is not dynamic enough to give strong incentives and price signals for emission reduction investments; a floating cap in relation to the most recent economic status of the EU might to be discussed;

---

[22]"Structural reform of the European carbon market", European Commission, 2013, downloaded 13.2.2013, http://ec.europa.eu/clima/policies/ets/reform/index_en.htm
[23]"Europe's $287bn carbon 'waste': UBS report", Sid Maher, The Australian, 2011, http://www.theaustralian.com.au/national-affairs/europes-287bn-carbon-waste-ubs-report/story-fn59niix-1226203068972 – downloaded 1 May 2013.

– The emission cap could be lowered, but its definition will always depend on political decision makers who do not have the interest to put additional burden on their industries especially in times of economic crisis;
– Carbon leakage still remains a problem;
– The renewables and the efficiency targets already reach most of the 2020 emission reduction targets, which questions the existence of the EU-ETS if no higher emission target will be established.

As a consequence, to make the EU-ETS a system which makes sense, a new binding 2020 target should be established to move the EU further beyond the already to be expected reductions in 2020 via the renewables and the efficiency targets. 30% emission reductions for 2020 are the new target discussed. With such a target and with a dedicated and constantly further reduced cap of the greenhouse gas emission allowances, the ETS market prices might reach levels, which really create incentives for investors to go for additional emission reductions.

4.6   CONCLUSIONS

With the climate package the European Union has introduced a set of targets for renewable energy, energy efficiency and reduction of climate gas emissions for the year 2020. Whilst the transposition and implementation are work in progress one can assume that the political frameworks on EU and national levels are in the first place able to deliver those 2020 targets. However, the devil lies in the details and a lack of ambition and completeness of legislation leads to the conclusion that all three targets are still under threat. More ambition, quality of legislation and willingness among decision makers will be needed to assure the real implementation of the climate package.

Renewables are in the year 2013 on track, but the expected rates of annual installations are going up towards 2020 and towards the 20% binding target and one has to fear that most Member States lack the willingness to really implement those targets by giving renewables markets the necessary push. Especially the negative news on renewables support in several EU Member States in the last years raise doubts about the way to go. More ambition and investors' security will be needed to have an overall share of renewables of 20% in 2020.

Energy efficiency is an even more tricky policy area. Efficiency is everywhere and not as easy to push as renewables. The 20% efficiency target for 2020 will not be reached from the perspective of 2013. The new energy efficiency directive surely has changed the situation and laid the ground for new policies which push target achievement from around 10% business-as-usual to expected 17% in 2020. However, more ambition and new policies will be needed to reach the 20% in 2020. A more comprehensive understanding how to implement a complex goal like the efficiency one will be needed and an effective set of policy packages will be needed to deliver.

The greenhouse gas emission reduction of 20% for 2020 lies in the first place in the success or failure of the European emission trading system (EU-ETS). Cap and trade is in theory an adequate measure to limit emissions and to find the most cost effective solutions to save the climate. However, the first phases of implementation of the ETS have shown that loopholes and systemic errors in the design of the EU-ETS lead to a complete failure of the system. Too many emissions allowances are on the market while the EU economy has reduced its activity during the economic crisis. This surplus of allowances leads to market prices, which are way too low to change the behaviour of market actors and to reduce greenhouse gas emissions. In addition, if one assumes that the renewables and efficiency targets are successful in 2020 while reducing emissions already on their own, the separate 20% emission reduction target with a complex tool like the EU-ETS does not make sense any more. It just does not deliver additional measurable emission savings. More and more voices can be heard which request at least a 30% greenhouse gas emissions target for 2020 in order to add ambition and incentives via the EU-ETS. Hope lies in this debate as well as in the new plans of EU institutions to heavily reduce the number of emission allowances which could lead to higher emission allowance prices and to incentives for market actors to further limit greenhouse gas emissions.

All in all, the success of the climate package and its three individual targets lies in the hand of political institutions on European, national and local levels. If decision makers see the opportunities in the implementation of those targets, success is possible: positive impacts for health and environment, economic growth, jobs, relief for public budgets and for Europe's independence of energy imports and capital exports. Strong ambition is the ground to deliver results. However, if Member States, its leaders and voters continue to see those measures as a burden, the 2020 targets will most probably be missed and EU loses major options to grow into an innovation based green and clean economy.

CHAPTER 5

Legal assessment of "discriminating market barriers" in national support systems

Markus Kahles & Thorsten Müller

5.1 INTRODUCTION

From the point of view of the European Law, one of the major objections to the national support schemes for electricity from renewable sources of the Member States has always been their restriction to domestic installations. This is because this characteristic of the national support schemes is in conflict with the guarantee of free movement of goods in accordance with Article 34 Treaty on the Functioning of the European Union (TFEU 2012), according to which quantity-related import limitations as well as all measures with an equivalent effect are prohibited between Member States. However, the national support schemes restrict cross-border electricity trading in several ways, which raises the question whether these restrictions are justified (refer to Chapter 6.2 of this book). In this regard, the 2009/28/EC Directive (Renewables Directive 2009) clarifies in accordance with the second subparagraph of Article 3(3) that the Member States are free to use discriminating support schemes (refer to Chapter 6.3 of this book). This legalisation by secondary law poses the question as to whether the support schemes of the Member States should still be measured directly by the free movement of goods (refer to Chapter 6.4 of this book).

5.2 BASIC CONFLICT: FREE MOVEMENT OF GOODS VERSUS NATIONAL SUPPORT SCHEMES

Article 34 of TFEU includes the guarantee of free movement of goods and reads as follows: *"Quantitative restrictions on imports and all measures having equivalent effect shall be prohibited between Member States".*

National support schemes restrict the cross-border electricity trading and thus the free movement of goods in two ways. For one, they restrict the trade with energy generated by conventional sources because this is pushed out of the market by the promotion of renewable energies. However, the restriction of the free movement of goods is justified on account of reasons pertaining to environmental protection and climate change as imperative reasons that are in the interest of the public (fundamental: ECJ 1979) and do not need to be discussed further. Here, the fact that in settled case law, the ECJ regards environmental protection as a mandatory matter of public interest should suffice [...] (ECJ 1985, ECJ 1988).

However, a greater justification is required for the limitation of electricity generation from renewable energies to domestic plants. The support schemes of the Member States are always restricted to such RES plants within their field of application[1] and this leads to a discrimination of the electricity generated from renewable energies from other Member States. In this connection, the PreussenElektra decision of the ECJ (ECJ 2001) was in favour of the Member States but

---

[1] Compare for Germany with § 2 no. 1 of the Renewable Energy Sources Act (EEG), "Erneuerbare-Energien-Gesetz" of 25 Oktober 2008 (BGBl. I S. 2074), as last amended by Article 1 Act of 17 August 2012 (BGBl. I S. 1754)", English version EEG 2012.

did not put all speculations to rest. The progressive achievement of the internal energy market by means of the internal market packages issued since the decision, namely the adoption of the internal market directives (Electricity Market Directive 2003 and Electricity Market Directive 2009) was quoted as an argument against this restrictive scope of the support schemes of the Member States[2]. Moreover, the impact of the support schemes on the internal market has grown along with the increase in the percentage of renewable energies being used for energy production in the Member States. At the end of the day, a decision has to be made as to whether priority should be given to the free market or to climate protection, if it is not possible to attain both these objectives. In view of the fact that an agreement on a uniform support scheme is not possible in the foreseeable future, the decision of the RES-Directive on the basic conflict between the free movement of goods and the national support schemes strengthened the freedom of the Member States to determine the design of their national support schemes for the time being (reference to Article 3.3). In this regard the Union legislator also acted in accordance with the European treaties.

5.3   LEGALISATION OF DISCRIMINATING SUPPORT SCHEMES BY THE RES-DIRECTIVE

The second subparagraph of Article 3(3) of the RES-Directive clearly states that the Member States can freely determine – they are only subject to the restrictions of the state aid regime – the extent to which they wish to promote energy from other Member States.

> *"Without prejudice to Articles 87 and 88 of the Treaty, Member States shall have the right to decide, (...), to which extent they support energy from renewable sources which is produced in a different Member State."*

Recital 25 of the RES-Directive provides insight into the rationale of the Union legislator for this far-reaching release:

> *"One important means to achieve the aim of this Directive is to guarantee the proper functioning of national support schemes, (. . .), in order to maintain investor confidence and allow Member States to design effective national measures for target compliance. This Directive aims at facilitating cross-border support of energy from renewable sources without affecting national support schemes. (. . .) In order to ensure the effectiveness of both measures of target compliance, i.e. national support schemes and cooperation mechanisms, it is essential that Member States are able to determine if and to what extent their national support schemes apply to energy from renewable sources produced in other Member States (. . .)."*

As made clear by the Union legislator in Recital 25 of the RES-Directive, the currently limited character of support schemes will continue to exist. These teleological considerations are also supported by historic and systematic factors (Müller 2009, p. 167). From the systematic point of view, the RES-Directive depends on stable support schemes in the Member States to attain its objective. As per Article 3(3) lit. a) of the RES-Directive, they are the main instrument used for attaining the binding objectives of the directive. The calculation method used for the percentage of renewable energies in the gross final consumption of energy of the relevant Member State is significantly related to the domestic production as per Article 5(3) and (4) in conjunction with Article 2 lit. f) of the RES-Directive. Energy quantities that have been generated in a different Member State or a third country can be taken into consideration only under the strict conditions of the cooperation mechanisms of Articles 6 to 11 of the RES-Directive. Moreover, the adherence

---

[2] Directive 2009/72/EC, as well as its predecessor Directive 2003/54/EC, aims at introducing common rules for the generation, transmission, distribution and supply of electricity. It also lays down universal service obligations and consumer rights, and clarifies competition requirements.

to the national support schemes as measures for attaining the objectives can also be explained with the help of the legislative history of the directive which was marked by the rejection by the Member States of the Commission's proposal to introduce Europe-wide trade with guarantees of origin, relevant for attaining the objectives (European Commission 2008a, Article 9).

5.4   ARTICLE 34 OF TFEU AS A TEST CRITERION FOR NATIONAL SUPPORT SCHEMES?

As per secondary legislation, discrimination between renewable energies from another Member State and renewable energies from local sources is allowed according to Art. 3 (3) subpara. 2 of the RES-Directive. At first, this provision seems surprising and raises the question as to whether national support schemes can still be checked against the free movement of goods according to Art. 34 TFEU. Secondary law thus could create a suspensory effect against the scope of the Fundamental Freedoms. But the Union legislator is only allowed to define such a provision for restricting the free movement of goods in the Member States regarding the support of renewable energies, if the provision is in accordance with the European treaties (refer to 6.4.1 for this). In this case the provision provides a suspensory effect for the Member States regarding the scope of Art. 34 TFEU (refer 6.4.2 for this).

5.4.1   *Compatibility of the second subparagraph of Article 3(3) of the RES-Directive with the primary law*

The second subparagraph of Article 3(3) of the RES-Directive includes the permission for discriminative support of renewable energies by the Member States. Thus, this provision contradicts the free movement of goods as per Article 34 TFEU. Whether the Union legislator is bound to the Fundamental Freedoms is a topic of jurisprudential debates, but must be affirmed in the result even though the ECJ often gives a larger scope for discretion to the Union legislator than to the Member States[3]. Even if a binding effect equal to that of the Member States is adopted for secondary law provisions that allow Member States to establish or adhere to discriminating measures (Barents 1990, p. 22 ff. und ECJ 1978), the second subparagraph of Article 3(3) of the RES-Directive is considered a proportional restriction of the free movement of goods for the purposes of environmental protection and tackling climate change (Müller 2009, p. 169 ff.).

Article 34 TFEU does not unconditionally exclude measures that have a negative effect on the internal market. In fact, the establishment of an internal market sometimes has to be subordinated to other objectives. However, another important concern along with the establishment of a uniform market is a high level of environmental protection and tackling climate change. The provisions of the second subparagraph of Article 3(3) of the RES-Directive attempt to attain exactly these objectives, as specified in Recital 25 of the RES-Directive. There can be no dispute about whether these objectives – effective development of renewable energies and tackling climate change – are justified by a legitimate purpose. The restriction option enables the Member States to attain their defined mandatory objectives. This provision is necessary as well as proportionate. Thus there are no crucial legal objections to the second subparagraph of Article 3(3) of the RES-Directive.

It should also be noted that this legal permission is part of the regulatory provisions that overall contribute to the harmonisation of laws for supporting renewable energies. Defining mandatory targets stops the Member States from stepping back from the support of renewable energies, if they want to avoid infringement proceedings (reference to Article 3.3). Giving priority to questions about grid access and dispatching installations as per Article 16(2) lit. b) and c) of the RES-Directive is now mandatory. Additionally the cooperation mechanisms as per Articles 6–11

---

[3]Only compare with ECJ 2004, Para. 57; ECJ 2001a, Para. 37; ECJ 1997, Para. 27; ECJ 1993, Para. 11; ECJ 1984, Para. 15.

of the RES-Directive also aim at improved joint support of renewable energies by the Member States (reference to Article 3.3). Thus the RES-Directive overall contributes to harmonisation. The ECJ has stated on several occasions that the Union legislator has a wide margin of discretion determining the progress of harmonisation, if the action taken by the Union legislator overall contributes to harmonisation (ECJ 1984a, Para. 20; ECJ 1991; ECJ 1996, Para. 25; ECJ 1984b). The legal permission given to the Member States by the second subparagraph of Article 3(3) of the RES-Directive is thus in accordance with the free movement of goods.

5.4.2   *Suspensory effect of the second subparagraph of Article 3(3) of the RES-Directive*

In principal, the secondary law that is in compliance with the primary law presents a suspensory effect such that the ECJ no longer judges measures of the Member States within the scope of secondary law directly against the primary law but only on the basis of the provisions of the secondary law act (ECJ 1992, Para. 21; ECJ 2003, Para. 52–54). This suspensory effect generally proves to be disadvantageous for the Member States as they can no longer rely on the primary law grounds to justify deviating measures. However, in the case of legal permissions such as the second subparagraph of Article 3(3) of the RES-Directive, this suspensory effect could lead to measures taken by the Member States no longer being checked directly on the basis of the primary law. To this extent, a *"Protective effect"* would replace the *"Suspensory effect"*. However, the ECJ case law is inconsistent here. A few dogmatic questionable decisions, in which the ECJ considered measures of the Member States to be illegal due to violation of the Fundamental Freedoms without considering existing secondary law (ECJ 2003a, Para. 26; ECJ 2006, Para. 45), are confronted with a large number of cases where the ECJ checked the measures of the Member States only on the basis of the applicable secondary law (ECJ 2004a, Para. 56 f.; ECJ 2003, Para. 52–54; ECJ 2003b, Para. 58; ECJ 1992, Para. 20 ff.). Only in very few cases the secondary law was classified as incompatible with the primary law due to violation of the Fundamental Freedoms (ECJ 1978, ECJ 2007). The first-mentioned decisions should not be followed dogmatically for reasons pertaining to the hierarchy of standards and the clarity on the application of law. Thus it can be assumed that the support schemes of the Member States due to the second subparagraph of Article 3(3) of the RES-Directive should no longer be checked directly on the basis of the free movement of goods as per Article 34 of TFEU.

5.5   SUMMARY AND PROGNOSIS

Even if it seems surprising at first, the seemingly insignificant provisions of the second subparagraph of Article 3(3) of the RES-Directive have a far-reaching significance. The Union legislator has decided for the time being to put aside the objective of establishing a completely harmonised internal market in the field of renewable energies promotion. In fact, the legislator has given priority to the effectiveness and efficiency of the provisions of the Member States in order to achieve the renewable energy targets. This decision by the union legislator can be justified by the purpose of environmental protection and thus be compatible with European law. With this, the legislator has started a learning process that is important for future legal developments and that allows for the comparison between various regulatory approaches in the individual Member States.

   The position of the national support schemes in the Union policy for promoting renewable energies will thus in particular depend on the time period after 2020. In this context, the Commission came up with its Energy Roadmap 2050 (European Commission 2011f) in December 2011 which was recently concretised by the Renewable Energy Strategy (European Commission 2012b). Here, the Energy Roadmap 2050 must be considered together with the *"Roadmap for moving to a competitive low carbon economy in 2050"* (European Commission 2011a) and the *"Roadmap to a Single European Transport Area"* (European Commission 2011d) which have already been published in 2011. With the Energy Roadmap 2050, the Commission already at this stage wants to trigger a process for the period after 2020 which will help the Union to further the decarbonisation process of energy supply. The objective is to reduce the greenhouse gas emissions in the energy sector by at least 80% in 2050. The electricity sector plays a decisive role in this restructuring of the energy sector. Clarity about the future energy strategy of the Union should be obtained at the earliest in order to ensure the necessary investment and planning security for investors, decision-makers and citizens. Various scenarios are presented for this that may lead to various structural changes of the energy system, depending on the type of energy production or the amount of energy savings the Commission focuses on (Matthes 2012, p. 51). Here, the Commission relies on a significant share of nuclear power for the future energy production and the use of the CCS technology (European Commission 2011f, p. 8). In an ideal scenario for renewable energies, this comes to a 75% of the power supply in the year 2050 with soaring power costs, whereas in the unfavourable scenarios, the share is 55–57% also with increasing power costs which will however drop greatly after 2030 (European Commission 2011f, p. 6 f.). However, the lack of transparency in the basic assumptions and the calculation methods, which would make it difficult to test the scenarios, is a point of critique (May 2012, p. 16). Moreover, presenting the energy roadmap for 2050 is seen as the prelude to the discussion on the binding objectives of the support schemes for renewable energies for the period after the RES-Directive of the years 2020 to 2030 (Fischer and Geden 2012, p. 3). Parts of the renewable energies sector consider the expansion plan of the Commission of 30% by 2030 to be too unambitious (May 2020, p. 16). On the one hand, the support schemes of the Member States will in principle face increasing harmonisation pressure in the future due to the increasing share of renewable energies, particularly in terms of electricity supply, and the resulting distortion of the internal market. On the other hand, it is essential to provide stable framework conditions for further development of renewable energies in order to secure investments. Thus, in the foreseeable future, the debate on harmonisation will be ruled by the discussion about which role the renewable energies should play in the future European energy mix and by the question about the national or harmonised approach to the support of renewable energies.

CHAPTER 6

Powerful national support systems versus Europe-wide harmonisation – assessment of competing and converging support instruments

Markus Kahles & Thorsten Müller

6.1 INTRODUCTION

"The time has come for energy policy to become truly European". This is the claim made by the European Commission with respect to energy-related issues (European Commission 2010, p. 4). However, it raises the question how such a European approach can be structured in a meaningful manner. Wanting to answer this question by saying that only a harmonised system can be European would be premature and would negate differences as well as existing market failures and other necessary learning processes. It is not only reactionary forces in the Member States who consider energy policies as an integral part of national interest and are therefore against harmonisation. Rather, important questions pertaining to the meaningfulness and structure of such a procedure are still unresolved.

The ambivalence between the desire for harmonisation on the one hand and the consolidation of the existing support schemes on the other is also reflected in the directive for promoting the use of energy from renewable sources (Renewables Directive 2009). According to Recital 25, the directive follows the objective of *"facilitating cross-border support of energy from renewable sources without affecting national support schemes"*. According to this, the Member States can themselves decide in accordance with the second subparagraph of Article 3(1) of the RES-Directive to which extent they want to promote the energy generated from renewable sources in another Member State. Processes are initiated within this legal framework, which during the course of the competition among systems and the mutual learning process enable a convergence of the support schemes of the Member States even without central legal harmonisation and will definitely improve the understanding of effective as well as efficient approaches for promoting electricity generation from renewable sources. Subsequent to a brief overview of the support schemes used in the Member States, which is meant to explain the spectrum of the approaches followed in the Union for promoting electricity generation from renewable energies (refer to 6.2 for this), basic considerations about the conflict between desired increased efficiencies of harmonisation and possible disadvantages of a premature harmonisation have to be made (refer to 6.3 for this). Based on this, a description is provided as to how the coexistence of different support schemes can be used to determine the most efficient manner of promoting renewable energies through competition among systems (refer to 6.4 for this). This competition among systems is supplemented by the instruments of intergovernmental cooperation included in the RES-Directive, which enable the Member States to mutually promote renewable energies (refer to 6.5 for this).

6.2 OVERVIEW OF THE SUPPORT SCHEMES IN THE MEMBER STATES

The Member States have established different forms of support for electricity generation from renewable energies; some of them were already established in the 1990s before the 2001/77/EC Directive was adopted. The most commonly used measures for promoting electricity generation from renewable energies include feed-in tariffs, feed-in premiums and quota obligations, whereas tender models are hardly used nowadays such that they are subsequently neglected (ECOFYS 2011, p. 28; European Commission 2011k, p. 9 f.).

6.2.1   Feed-in tariffs

Different versions of feed-in tariffs are used by most of the Member States and distinguish themselves by the fact that the producers of electricity from renewable energies receive a fixed compensation for a particular period of time from the transmission system operators or energy supply companies for the energy fed into the transmission system. The amount of the compensation varies mostly depending on the type of the renewable energy source used. Due to the specified regulatory compensation rates this approach is also referred to as a price control system. At present, 21 Member States use such feed-in tariffs (European Commission 2011k, p. 6).

The German Renewable Energy Sources Act (EEG) is the most prominent among this type of promotion and is often considered to be an example for the European development (OPTRES 2007, p. 18). The core elements of this law are

- the priority connections of installations for generating electricity from renewable energies to the supply network for general electricity supply (sect. 2 No. 1 EEG),
- the priority purchase, transmission, distribution and payment for such electricity by the system operators as well as the payment of premiums for integrating this electricity into the electricity supply system (section 2 No. 2 EEG) and
- the nationwide equalisation scheme for the quantity of electricity purchased and paid for (section 2 No. 3 EEG).

Conventional electricity will be pushed out of the market due to the priority given to electricity produced from renewable energies in combination with its financial support.

6.2.2   Feed-in premium

The so-called feed-in premium systems are a variant of the feed-in tariffs. These systems also deal with price regulations. However, unlike the feed-in tariffs, the investment incentive is not based on a comprehensive and guaranteed payment. Rather, feed-in premiums force the producers of electricity from renewable energies to offer and sell on the market. In addition to the sales profit, feed-in premium systems then ensure a premium which is supposed to compensate for the difference between the achieved market price and the increased cost price plus a reasonable profit margin (European Commission 2011k, p. 5). The Spanish support scheme for renewable energies is considered as a prototype of this approach (ECOFYS 2011, p. 30). Since 01.01.2012, the so-called *"market premium"* in accordance with section 33b No. 1 of EEG can now also be selected as compensation model in the German EEG, as an alternative to the fixed feed-in compensation.

6.2.3   Quota obligations

Quota obligations indicate the regulatory alternative approach to feed-in tariffs and feed-in premiums. Electricity suppliers or consumers are under obligation within the scope of the quota obligations to cover a specific percentage of their electricity requirement from renewable energies. Hence, instruments for quantity control are referred to as well. Certificates which must be acquired by the obligated parties serve as a proof of fulfilment of this rate. Owing to this structure, rate regulations are considered to be more "market based" than feed-in tariffs as the market mainly decides about the price and thus also about the type of regenerative energy production that should be used (European Commission 2005, p. 5). In order to avoid that the legally required share of renewable energies comes only from the sources having the lowest cost price, some quantity control approaches provide individual quotas for some technologies (*"banding"*). However, past experiences have shown that in the end this approach is less effective than feed-in tariffs

(Klessmann 2011, p. 30). In addition, the determination of the rates leads to a restriction of the development effect to the legally determined extent. Another weak point is sanctions, which are often not efficient enough, in case the quantity specifications are not met (OPTRES, 2007, p. 131).

6.2.4 *Differences and similarities*

The different approaches for promoting renewable energies by the Member States presented above are distinguished by the fact that they limit their promotion effect, for instance in Germany in accordance with section 2 No. 1 EEG, to the electricity generated domestically from renewable energies. Thus, from the legal point of view, all the support schemes conflict with the free movement of goods according to Article 34 of TFEU (Treaty on the Functioning of the European Union). The restriction on the free movement of goods resulting from the promotion of renewable energies by the Member States must however be justified mainly for reasons of environmental protection and climate change. The RES-Directive now specifically allows the different treatment of foreign electricity from renewable energies in accordance with the second subparagraph of Article 3(3) of the RES-Directive (see Chapter 5 of this book).

In the past the European Commission has considered quota obligations as *"closer to the market"* and hence as generally more compatible with the state aid legislation in accordance with Article 107 of TFEU (European Commission 2005, p. 5). Since no state resources are essentially included in case of quota obligations, they are not regularly subject to the state aid check. However, feed-in tariffs can also be developed such that the state aid rules do not apply (ECJ 2001). In addition, the promotion of renewable energies can mainly be considered as consistent with the requirements of the internal market and the state aid rules in accordance with Article 107(3) lit. c) of TFEU (Community Guidelines 2008).

On the other hand, a final assessment of the economic advantages and disadvantages with respect to the effectiveness for promoting electricity generation from renewable energies cannot and should not be carried out owing to the jurisprudential focus of this article. However, it can be concluded from studies that have been conducted that a Europe-wide quota obligation would probably not result in the desired efficiency effects (REShaping 2010).

The development success that has already been achieved with electricity generation from renewable energies in countries which use feed-in tariffs, advocate the use of such a system. This way, a greater support for different technologies seems possible whereas quota obligations tend to prefer the technology that is most efficient according to costs and benefits (ECOFYS 2011, p. 104). However, in a stage in which there is a lot of scope for the technological development of renewable energies and other forms of market failure have to be ascertained, a too early commitment by the market to a few *"cheap"* technologies seems a rather undesirable effect. With a wide support by means of fixed feed-in tariffs, the required investment safety can be created on the contrary so as to provide the option to a large number of promising technologies to compete with conventional energy producers in the foreseeable future. Undesired windfall profits on part of the plant operators and economic inefficiencies can and must be counteracted within the feed-in systems, for instance, by means of a technology-specific diversification which reflects the development status of the respective technology or a percentage reduction of the feed-in tariffs (degression).

For a long time now, the European Commission and established companies in the electricity industry have favoured quota obligations (EWI 2010; European Commission 2005 p. 5). In the meantime, a suggestion was also made to introduce a Europe-wide certificate trading (European Commission 2008a, Recital 19). However, at least the Commission seems to have deviated from its preference for rate models in the meanwhile; in fact, it seems to favour a regulation with feed-in premiums (European Commission 2011i, p. 11; European Commission 2011k, p. 7).

6.3 HARMONISATION OF THE SUPPORT SCHEME AS AN ALTERNATIVE?

The coexistence of different forms of support in the European Union and their scope regularly restricted to the respective Member State has for a long time now generated demands for a harmonisation on the basis of a uniform instrument. The support schemes of the Member States mainly seem to be exposed to harmonisation efforts of two sides. On the one hand, a high justification pressure for explaining the unequal treatment of foreign electricity from renewable energies emanates from the application of free movement of goods to national support schemes. This process is also referred to as *"negative integration"*. On the other hand, the introduction of a Europe-wide support scheme during the course of the "positive integration" is requested by the Union legislator.

The process of *"negative integration"* during the course of the jurisprudence of the European Court of Justice (ECJ) has not directly affected the promotion of renewable energies by the Member States till now, mainly because of its *PreussenElektra* decision (ECJ 2001) in which the German support scheme was assessed to be compatible with the fundamental freedoms. In this decision, the ECJ regarded the German support scheme for generating electricity from renewable energies as a justified restriction of the free movement of goods for reasons of environmental protection. This view has now also been introduced into the European secondary law. The second subparagraph of Article 3(3) of the RES-Directive allows the Member States to decide themselves about the extent to which they want to promote electricity from renewable energies from other Member States (see Chapter 5).

A legal harmonisation in the course of the *"negative integration"* must thus not be expected in the foreseeable future. This type of harmonisation would also only exclude individual regulations of the Member States that are not in conformity with the internal market. However, negative integration from the European perspective does not create a uniform legislation which is consistent within itself. Therefore, the supporters of a *"positive integration"* in the course of law-making by means of a harmonised support scheme hope for a cost-benefit allocation which is more efficient compared to separated markets and an improvement in the cross-border electricity trade. However, these expected gains in efficiency conflict with the still existing disagreement regarding the most efficient method of promoting renewable energies. Therefore, an outline will be provided as to how the status quo of the coexistence of different support schemes can be considered as a competition among systems and can be used as an integration method.

6.4   THE COMPETITION AMONG SUPPORT SCHEMES IN THE UNION

The development processes of the current RES-Directive and its predecessor, Directive 2001/77/EC, clearly show that till now there is no consensus on the question as to which support scheme is the best one for electricity generation from renewable energies in the European Union. There is consensus only about the necessity to promote renewable energies. Hence, the commitment to a specific Europe-wide support scheme seems too early. The RES-Directive thus selects an approach which preserves sovereignty by determining binding national targets, but leaving the choice of the means to achieve these objectives to the Member States as far as possible. The RES-Directive tries to make use of the coexistence of the different support schemes, but does not establish any pure competition among systems. Rather, a regulated competition among systems can be referred to. This way, possible negative consequences such as a *"race to the bottom"* can be avoided and a process of mutual learning can be enabled.

6.4.1   *Advantages and disadvantages of competition among systems*

The notion of competition among systems mainly comes from economic considerations of the *"Economic Theory"* (Posner 2007). The so-called *"economic principle"*, according to which individuals or companies in the free market, using the resources available to them, always strive to achieve the best possible result or achieve a particular result as efficient as possible , is applied to the legal systems created by the individual states. Here it is assumed that only those systems will be implemented in the long run which prove to have an edge over other competitors with certain economic factors such as workplaces and investments.

It is also assumed that in case of disagreement or uncertainty regarding the correct regulatory approach, this uncertainty can be resolved in a competition among the legal systems. If generally binding rules are set centrally by the Union legislator, a parallel experimental process with different simultaneously existing rules is no longer possible. Especially in cases where there is uncertainty or disagreement about the correct procedure, it would therefore be advisable not to rely on the central harmonising regulation, but to exhaust the innovative potential of different attempts at a solution by means of mutual learning (fundamental: Hayek 1968).

Particularly in the company law, making complete use of different regulations in the Member States, based on the freedom of establishment, is interpreted as competition among systems (ECJ 1997a). If these notions are applied to the promotion of renewable energies, the innovation potential of different legal systems can thus be exhausted in a competition among support schemes which, in terms of a *"bottom-up"* approach, which makes it possible to find the best method for promoting renewable energies.

The basic requirement for a functioning competition among systems is the mobility of economic factors, in particular, of workers, enterprises and capital so that any legal system can be selected from those available. This is mainly ensured within the European Union through the fundamental freedoms, even in case of restriction of the support schemes of the Member States for renewable energies to domestic installations. This way, companies, employees and investors can freely select between the different legal systems and choose the one which offers the best conditions for their respective requirements (ECJ 1997a, Para 20ff). In order to be able to find out which legal system is best suited to the respective requirements of the market participant, it is important to know about the advantages and disadvantages of the respective legal systems which are available for selection in the European Union. The more easily such information is available, the greater is the mobility because high information costs basically have a limiting effect on the factor mobility.

Different functions and the resulting advantages are attributed to the competition among systems as compared to the central harmonisation (Streit 1996; Reich 1992). This is a controlling function on the one hand. This results from the assumption that in the case of a central regulation there is a risk that the comparability of a legal solution selected once becomes inapplicable due to lack of competition. Error correction can become more difficult as a result. This is because, a uniform legal framework, as a product of negotiation, can often turn out to be just an *"average solution"* in which the different preferences cancel each other out. In addition, local knowledge can be made more productive with a decentral regulation.

Decentral regulation can also counteract the so-called *"rent-seeking"* problem. This phenomenon means that the more central the legal decisions are made, the greater the likelihood is that the decision makers will lose track of the general interest due to the influence of interest groups. Smaller decision-making units on the other hand limit the power of individual decision makers and increase the comparability between the different regulations systems, which makes it easier to recognise faults in the design.

Owing to these features, the competition among systems, especially in transitional phases, (for example, before secondary law is issued) can be used for observing the different national approaches over a longer period in order to then utilise the methods tried and tested in the Member States. A premature harmonisation in terms of *"levelling the playing field"* is often not advisable in view of the lack of knowledge regarding the appropriate method.

However, applying the theory of competition among systems to the different regulation systems in the Member States not only has advantages. For example, the risk of a so-called *"race to the bottom"* is especially inherent in competition among systems, also known as the *"Delaware* effect" (Streit 1996; Reich 1992). In theory, such a *race to the bottom* results in the competing states reciprocally underbidding during the deregulation and lowering of the minimum requirements so as to create the most liberal investment conditions and as a result they neglect the protection of certain environmental, health or social standards.

The risk of transferring *external effects* is closely linked to this problem, mainly in case of cross-border matters, such as the environmental and climate policy. Here the basic problem of resources of common interest crops up at the intergovernmental level as well; there are hardly any incentives for individuals to contribute to the upkeep (so-called externalisation) of resources of common interest. The climate, specifically a particular CO_2 percentage in the atmosphere, is such a resource of common interest, whose pollution is however generally not reflected in the success or failure of a particular national economic or energy policy which has a restricted scope. On the other hand, a non-ecological economic policy can also be reflected in the high growth rates of a single country while everyone has to pay the price for environmental and climate pollution. Cooperation between the affected countries can prevent that externalities are shifted and thus inefficiencies between the different regulation systems are exploited to the benefit of one country and at the expense of all countries.

Finally, a reference is made to the fact that an apparent advantage of a system can arise solely due to the dominance of individual systems owing to the economies of scale or economic power. However, this is not the final result of a fair competition among systems, but results from an already predetermined distortion of the competition (Dreher 1999).

6.4.2 Avoiding disadvantages by means of binding targets

Directive 2009/28/EC (Renewables Directive 2009) uses the method of competition among systems linked to the elements of minimum harmonisation and cooperation as an alternative to complete harmonisation. Despite individual harmonising aspects, for instance, with respect to grid connection, the RES-Directive does not establish a uniform support scheme for renewable energies in Europe, but allows the different support schemes of the Member States to exist; in fact, it strengthens the legislatory discretion of the Member States.

In order to counteract the above described detrimental sides of a competition among the support approaches and to use the positive aspects at the same time, different measures are taken: main importance is given to the binding objectives determined for every country in the directive for a proportion of renewable energies to be reached in the total gross final consumption of energy. The Union wants to reach a minimum percentage of 20% renewable energies in the gross final energy consumption in the year 2020 in accordance with Article 3(1) of the RES-Directive. For this purpose, individual binding targets are set for every Member State in the form of a specific percentage of renewable energies in the gross final energy consumption. In accordance with Annexe I of the RES-Directive, these range from a percentage of 10% for Malta to up to 49% for Sweden. The objective for Germany is 18%. The percentage of electricity or heating/cooling from renewable energies in the respective final energy consumption of the Member State is thus not specified. For transport however, Article 3(4) of the RES-Directive bindingly prescribes for each Member State that in the year 2020 the share of energy from renewable sources in all forms of transport has to be at least 10% of its final energy consumption in transport.

Article 5(2) of the RES-Directive states a hardship clause where a Member State is, as an exception, allowed to deviate from its binding objective in case of force majeure. The Commission decides whether a case of force majeure was proven and, if necessary, revises the objective downwards. If a Member State falls short of its objective, without being allowed to rely on force majeure, no special sanctions are provided for in the RES-Directive. However, the Commission has the option of initiating a treaty infringement procedure according to Article 258 TFEU on the basis of the liability of the objectives. Such proceedings can nevertheless not be initiated if a Member State negatively deviates from its indicative trajectory according to Annexe I Part B of the RES-Directive since the intermediate targets are not binding (Müller 2009, p. 161). However, in case of a deviation from their indicative trajectory according to Article 4(4) of the RES-Directive, the Member States must submit a revised national action plan in which they must state the measures they intend to use in order to re-comply with the trajectory within an appropriate time span.

Finally, the competition among systems should be influenced in a way that ensures that all Member States implement effective support schemes. This prevents the possible temptation of avoiding, on grounds of economic considerations, a rise in the electricity price owing to promotion of renewable energies and thus attempting to secure a local advantage over other Member States.

The binding targets thus significantly contribute towards ensuring that there is no *race to the bottom*.

Rather, a *race to the top* can be initiated by means of ambitious binding targets. Advantages can be searched for by selecting the most efficient possible method for promoting renewable energies in order to achieve the respective national targets of the directive. Economic advantages beckon such as technological leadership and job creation and, in case of overfulfillment, possibly even a profitable utilisation of the cooperation mechanisms according to Articles 6–11 of the RES-Directive.

6.4.3   *Establishment of an institutional framework for information exchange*

In addition to the binding targets, the effects and consequences of the individual forms of support of the Member States are subject to a comprehensive evaluation and transparency process. To this end, the Commission monitors the improvements in the measures for promoting renewable energies according to Article 23 of the RES-Directive. According to Article 23(8) lit. c) of the RES-Directive, the Commission shall submit a report until 31st of December 2014 which, among other things, contains an assessment of the implementation of the RES-Directive especially with reference to the cooperation mechanisms and with reference to reaching the 20% objective. The Commission can suggest measures on the basis of this report, which however according to the second bullet point of Article 23(8) lit. c) of the RES-Directive may neither result in a change to the 20% objective nor are allowed to affect the control of the Member States over their national support schemes or measures that are already taken with respect to the cooperation mechanisms.

In addition to the reporting obligations of the Member States towards the Commission and the release of reports by the Commission based on this, in accordance with Article 23(3) of the RES-Directive, numerous other initiatives follow the institutional establishment of the exchange of information between the Member States with respect to promoting electricity generation from renewable energies. The so-called "*transparency platform*" is set up and supported by the Commission[1]. Article 24 of the RES-Directive requires the setting up of such an online transparency platform. According to Article 24(1) of the RES-Directive, this should serve to increase transparency and simplify and promote the collaboration between the Member States, especially with reference to statistical transfers in accordance with Article 6 and joint projects according to Articles 7 and 9. In addition, the "*res-legal*" information platform was set up on behalf of the Commission which makes the support schemes of the Member States visible, updates and explains them.[2]

A common information platform for the Member States is the so-called "*Concerted Action Renewable Sources Directive*" (CA-RES).[3] CA-RES is supposed to enable a structured and strictly confidential dialogue between the Member States for the purposes of implementing the 2009/28/EC RES-Directive. In addition, the process of mutual learning should be strengthened and a dialogue regarding common approaches for implementing the RES-Directive must be achieved. CA-RES is supported by the Intelligent Energy Europe (IEE) programme of the EU and coordinated by the Austrian energy agency. The "*International Feed-in Cooperation (IFIC)*[4]" forms another type of cooperation between the Member States at the political level, within which Member States, which use feed-in tariffs for promoting renewable energies, have united. Members include Germany, Spain, Slovenia and lately Greece also. Most other MS regularly participate in the meetings and discussions.

In addition, the Commission regularly initiates different economic studies (OPTRES 2007; REShaping 2011; ECOFYS 2011) within the scope of different support programmes, e.g. the

---

[1] http://ec.europa.eu/energy/renewables/transparency_platform/transparency_platform_en.htm
[2] http://www.res-legal.de
[3] http://www.ca-res.eu
[4] http://www.feed-in-cooperation.org

before mentioned *Intelligent Energy Europe* – Programme or the *Sustainable Energy Europe*-Programme so as to evaluate the existing support schemes and thus identify the best method for promoting electricity generation from renewable energies (European Commission 2011i, p. 14).

6.5   PROCESS OF CONVERGENCE OF THE SUPPORT SCHEMES?

In particular in the promotion of electricity generation from renewable energies, it is observed that the different support schemes of the Member States do not coexist independently, but a process of mutual learning has definitely been implemented that results in elimination of some and probably more and more differences between the different support schemes. A few Member States alternate between quota and feed-in schemes, such as Great Britain, which shows that a lot is happening in the field of instruments. Even in Germany, the pioneer of feed-in tariffs, an attempt is being made from 01.01.2012 to make electricity generation from renewable energies less dependent on the fixed feed-in compensation and thus enter the electricity market by introducing an optional market premium which can be selected as an alternative to the fixed feed-in tariff. The Commission had also noticed a process of convergence of the support schemes in its working paper to its communication *"Energy 2020"*; however, the Commission thinks that the progress is too slow and uncoordinated (European Commission 2011k, p. 10 f.).

However, with the cooperation mechanisms according to Articles 6–11 of the RES-Directive, the RES-Directive offers instruments unused so far with which Member States can jointly promote energy and especially electricity generation from renewable energies and can simultaneously contribute towards fulfilling their binding objectives.

6.5.1   *Statistical transfer*

According to Article 6 of the RES-Directive, the Member States can agree on the statistical transfer of a specified amount of energy from renewable sources from one Member State to another Member State for one time or for a longer period. The transferred energy amount is then deducted from the target compliance of the transferring Member State and allocated to the target compliance of the recipient state. It is thus made possible for Member States, which overfulfill their binding predetermined objectives, to deliver amounts of energy to other Member States, which do not want to or cannot reach their objectives with national development of renewable energies. The transfer is done purely statistically; the electricity is not actually supplied (European Commission 2011k, p. 7). Subparagraph 2 of Article 6(1) of the RES-Directive contains an important restriction when carrying out the statistical transfer. According to this, statistical transfer may not be carried out if the target compliance of the transferring Member State will be impaired in the process. Article 6(2) and (3) of the RES-Directive contain the notification obligations towards the Commission.

Within the scope of statistical transfer, the cooperation between the Member States is restricted to pure shifting of amounts of energy *"on the balance sheet"* and represents the simplest form of cooperation mechanisms in this respect. Private companies cannot be included directly as neither the guarantees of origin according to Article 15 of the RES-Directive nor any other tradable RES certificates (e.g. RECS) have an influence on the target compliance of a Member State. The target compliance is calculated according to Article 5 of the RES-Directive directly on the basis of the share of RES electricity in the gross final consumption of energy in the respective Member State. Private companies thus cannot sell a particular share of the RES electricity produced by them, guaranteed by certificates, to authorities in other Member States and in this way increase their RES balance statistically. The statistical transfer according to Article 6 of the RES-Directive thus significantly differs in its functioning from the international emission trading or the original plans for introducing a Europe-wide certificate trade for electricity from renewable energies.

The effect of promoting the internal market lies merely in the possibility that a market for statistical amounts of energy from renewable sources could be established *among the Member States*. However, this would only result in a very indirect harmonisation of the different forms of support of the Member States because the statistical transfer does not require any agreement among the Member States regarding the method of promoting renewable energies. At best, it leads to some competition with respect to efficiency. In the first place, the statistical transfer can help to safeguard those Member States in which a target breach is most likely as the year 2020 draws closer (REShaping 2010a).

6.5.2   *Joint projects*

As another form of intergovernmental cooperation, the RES-Directive, according to its Articles 7–9, introduces joint projects between Member States or between Member States and third countries.

6.5.2.1   *Joint projects between Member States*

According to Article 7(1) of the RES-Directive, two or more Member States can work together within the scope of joint projects for generating electricity, heating or cooling from renewable energy sources. In contrast to statistical transfers, the collaboration can also include private operators. This means that private plant operators can be supported within the scope of joint projects or, alternatively, also that private project planners can identify suitable projects (European Commission 2011k, p. 8). A joint project must be a newly constructed installation that became operational after 25 June 2009 or an installation with increased capacity that was refurbished after that date. Not only the installation can be co-financed by several Member States but the relevant infrastructure as well (European Commission 2011k, p. 8). The statistical amounts of energy from renewable sources resulting from such a joint project are distributed proportionately among the Member States according to Article 7(2) of the RES-Directive in accordance with the agreement between them and thus serve the target compliancy. The actually produced energy must not follow this virtual distribution pattern, but can be independently supplied by the private plant operator to his/her consumers.

Article 7(3) of the RES-Directive contains various notification requirements towards the Commission with respect to the type of installation, the distribution of the statistical amounts of energy and the time period of this agreement. Article 8 of the RES-Directive contains other notification obligations towards the Commission which have to be carried out during the project in order to fulfil the target of the respective Member State. Joint projects according to Article 7 of the RES-Directive show similarities with the "Projects of joint implementation" according to Article 6 of the Kyoto protocol. Thus it can be argued, that the Projects of joint implementation served as an example for the joint projects (Müller 2009, p. 158).

The question with respect to the details of the joint promotion of these projects can be answered differently within the scope of the intergovernmental agreements. A harmonisation of the support schemes is thus not inevitably linked. Compared to statistical transfers, joint projects between Member States according to Article 7 of the RES-Directive however require at least an agreement about certain aspects of the joint promotion of renewable energies between at least two Member States. This effect can be restricted to a few certain aspects as the cooperation can refer only to separately defined projects or an individual project and also the support schemes in the Member States need not necessarily be used for financing such projects. Instead, it is probable that two or more Member States agree upon a special support mechanism parallel to the existing support scheme specially tailored to a specific project and a particular technology and that they, for instance, determine the project operator by a tender procedure (REShaping 2010a).

However, it may not be underestimated that within the scope of joint projects the Member States essentially make an agreement regarding the joint promotion of electricity generation (or heating and cooling) from renewable energies. Even if this form of cooperation is initially restricted only to individual projects and in this respect the implementation of joint projects also does not necessarily require that different support schemes in the Member States are changed and harmonised with each other, the decision makers in the respective Member States however demand an intense engagement with the existing support schemes of the potential partner states before the implementation of a joint project in order to determine the best framework conditions for the respective joint project. Mutual learning processes can be initiated from this.

6.5.2.2   Joint projects with third countries

In addition to the joint projects between Member States, the RES-Directive, according to Article 9, also provides the option to the Member States to contribute to their target compliance by promoting the generation of electricity from renewable sources by means of joint projects with third countries. The cooperation can include private operators within the scope of Article 9 Projects as well. However, the electricity generated from renewable energies in joint projects outside the Union can only be allocated to the target compliancy of the involved Member States if, according to Article 9(2) lit. a) to c) of the RES-Directive, the electricity is consumed in the Community, the installation concerned was rebuilt or refurbished to increase its capacity after the RES-Directive came into effect and no support from a support scheme of a third country other than investment aid is granted to the installation.

Thus, for joint projects with third countries, the special prerequisite is that the amount of electricity generated in such projects is actually imported into the Union and consumed there. As proof of such an import, Article 9(2) lit. a) of the RES-Directive prescribes that firstly an amount of electricity equivalent to the electricity accounted for was firmly nominated by all the responsible transmission network operators in the country of origin, in the country of destination and, if relevant, in each third country of transit to the allocated interconnection capacity; secondly, an amount of electricity equivalent to the electricity accounted for was firmly registered in the schedule of balance by the responsible transmission system operator on the Community side of an interconnector and thirdly, that the nominated capacity and the generation of electricity from renewable energy sources by the installation operated within the scope of a joint project refer to the same time period. Article 10 of the RES-Directive contains other notification obligations towards the Commission.

Like joint projects between Member States according to Article 7 of the RES-Directive, elements of the flexible mechanisms according to the Kyoto protocol can also be detected in case of joint projects with third countries according to Article 9 of the RES-Directive. In this case, the mechanisms at question are the Clean Development Mechanisms according to Article 12 of the Kyoto protocol (Müller 2009, p. 158 f.).

Article 9 of the RES-Directive integrates into the energy policy of the Union, in which international collaboration with third countries plays an important role. For instance, this includes promoting cooperation in renewable energies projects with the countries of the south Mediterranean area (*Mediterranean Solar Plan*). Especially the start of pilot installations in the 2011–2012 period is intended here (European Commission 2011e, p. 6). Furthermore the *DESERTEC* initiative is an example for a private initiative, which mainly encourages the implementation of solar projects in North Africa[5]. For such projects, the use of Article 9 of the RES-Directive represents another possibility of receiving financial support from the Member States; in return, such projects can contribute to the target compliancy of the Member States by using Article 9 of the RES-Directive.

6.5.3   Joint support schemes

Article 11(1) of the RES-Directive provides the option that two or more Member States can voluntarily decide to join or partly coordinate their national support schemes. The energy volumes generated from renewable sources using such a support scheme can either be distributed among the involved Member States according to Article 11(1) lit. a) of the RES-Directive by means of a statistical transfer or, according to Article 11 (1) lit. b) of the RES-Directive, the energy volumes can be allocated to the target compliancy according to one of the distribution rules set up by the involved Member States.

---

[5] http://www.dii-eumena.com/

Out of the three cooperation mechanisms, joint support schemes thus require the greatest extent of cooperation between the involved Member States and, in this respect, represent a strong form of intergovernmental coordination. However, the joint support schemes need not refer to all types of renewable energies, but can also be agreed upon only with respect to individual technologies or regions. However, there are still questions regarding the delimitation between joint projects according to Article 7 of the RES-Directive and joint support schemes. The difference to the promotion of a variety of joint projects must, in case of such a specified joint support scheme, be seen in the fact that the support scheme is generally determined in advance for an initially undefined number of installations (although in such cases, a quantitative upper limit for the promoted energy volume can also be thought of in the form of a "*cap*") and is not specially prepared for one or more projects which are defined more precisely. The energy volumes can also be distributed differently among the Member States for every joint project, which is not possible in case of a previously determined distribution rule according to Article 11(1) lit. b) of the RES-Directive.

On the condition that Norway implements the RES-Directive as a non-EU member, the *Joint Certificate Scheme* agreed upon between Norway and Sweden in 2011 can be valid as the first application case of a joint support scheme according to Article 11 of the RES-Directive (platts 2011, RES4less 2011). However, the negotiations regarding this were started in 2001 and had already failed once in 2006 due to financing problems (REShaping 2010a). This example shows how difficult and tedious the agreement regarding a joint support scheme can be, even in case of similar starting positions. It also clearly demonstrates that joint support schemes can be as a first step used particularly to merge systems which anyway have similar features. On the one hand, this contributes to rectifying the existing differences between similar systems and helps the internal market in this respect. However, on the other hand, there is a risk of block formation along the already existing division between quota and feed-in systems. A real "*breakthrough*" for the convergence of national support schemes would be ensured only if several Member States with different approaches to the support of energy from renewable sources can agree upon a joint support scheme. The joint support schemes according to Article 11 of the RES-Directive thus come as close as possible to the ideas of a Union-wide harmonised support scheme, but require a high degree of intergovernmental cooperation.

6.6 SUMMARY

The national support schemes of the Member States, at least in the foreseeable future, continue to be in a stress ratio to the objective of achieving the internal energy market and the harmonisation attempts of the Commission. This article has however shown that the coexistence of functioning support schemes need not primarily be interpreted as an obstacle to a uniform European promotion policy. Given the uncertainty regarding the correct design of a Europe-wide support scheme, the different approaches in the Member States can be used in competition and in a process of mutual learning to achieve a harmonisation with a "*best-practice*" method for promoting renewable energies. This may, in the long run and regarding the aspect of efficiency, not replace a harmonisation or not even lead in a kind of natural process to a support scheme "*from the bottom-up*", but it can absolutely prevent a premature commitment to a uniform support scheme and protect properly functioning national regulations from a hasty harmonisation. For this purpose, the RES-Directive, with the commitment to binding targets linked with great discretion for Member States and a simultaneous evaluation and transparency process, preserves the sovereignty of the Member States and also corresponds to the original ideas of the directive as legal method for integration. In the course of intergovernmental cooperation, the cooperation mechanisms of the RES-Directive offer Member States opportunities for developing a coordinated promotion of renewable energies themselves instead of waiting on future suggestions by the Commission. In view of the next round of the harmonisation debate to be expected, the Member States should not lose out on this opportunity to already achieve a convergence of renewable energy support in accordance with the preferences of their respective support schemes.

CHAPTER 7

Internal energy market – implementation still pending

Rainer Hinrichs-Rahlwes

7.1 INTRODUCTION: ENVIRONMENT, ENERGY, FREE MOVEMENT OF GOODS

One of the most ambitious endeavours of the European Union is the establishment of a single European market, transforming the whole Union into an internal market for producers and consumers alike: Article 26 (2) of the Lisbon Treaty – in continuation of earlier versions of the European treaties – outlines the objectives: *"The internal market shall comprise an area without internal frontiers in which the free movement of goods, persons, services and capital is ensured in accordance with the provisions of the Treaties"*. The objective of an internal market is complemented by specific regulations for the free movement of goods (Articles 28 and 29) and of people in the whole Union.

In the energy sector barriers for a single market have been very high from the beginning. Specific legislation was therefore designed, discussed and enacted, taking into account the various and contradictory national energy sectors, where deeply integrated incumbent monopolies or oligopolies have played and are still playing dominant roles, making it very difficult to establish functioning markets and even more so to *"liberalise"* them and to open them up for new entrants, for competitors striving for market shares.

The energy mix, the organisation of the sector and related policies was and still is considered to be a specific prerogative of national governments and parliaments, with or without direct reference to the subsidiarity principle. All decisions of the European Union regarding support for renewable energy so far were therefore legally based on the environmental competence, giving the Union the right to decide – with a qualified majority in Council and Parliament – about targets and policies, just as it was the case when the climate and energy package for 2020 was discussed, agreed and put in place.

Only in 2010, with the Lisbon Treaty entering into force, the European Union acquired a partial energy competence as laid out in Article 194, paragraph (1): *"In the context of the establishment and functioning of the internal market and with regard for the need to preserve and improve the environment, Union policy on energy shall aim, in a spirit of solidarity between Member States, to (a) ensure the functioning of the energy market; (b) ensure security of energy supply in the Union; (c) promote energy efficiency and energy saving and the development of new and renewable forms of energy; and (d) promote the interconnection of energy networks"*. However, following Member States' reservations, paragraph (2) underlines that this *"shall not affect a Member State's right to determine the conditions for exploiting its energy resources, its choice between different energy sources and the general structure of its energy supply without prejudice to Article 192(2)(c)"* According to Article 192 (2) the EU can – though this requires unanimity – decide about *"measures significantly affecting a Member State's choice between different energy sources and the general structure of its energy supply"*.

The energy competence of the European Union adds to the environmental competence as laid out in Article 191. An overlap of energy and environmental competences remains, in particular when it comes to *"prudent and rational utilisation of natural resources"* or *"promoting measures at international level to deal with regional or worldwide environmental problems, and in particular combatting climate change"*. Future discussions and decisions about common European objectives for sustainable energy policies and climate protection measures will have to build on this legal framework, which leaves some room for interpretation and thus for rulings of the European Court of Justice.

On this background, enacting and implementing the Internal Energy Market is a specific challenge for governments, legislators, regulators and market actors. The challenge begins well before the actual enforcement of a single European market. It begins with defining the targets and the related principles. The energy sector was (and to a certain extent still is) dominated by incumbent monopolies or at least oligopolies, rather than functioning as a market. Therefore, the task was not only to harmonise regulations between Member States, but to establish the basic preconditions of markets within the Member States: transparency of regulations and cost assignments, competition between different market players, choice for consumers between different gas or electricity suppliers. Particularly, where infrastructure and energy production are owned by the same companies, new market entrants do not have a fair chance to enter the market. Consequently, it is particularly challenging for independent producers of energy – even more so from renewable sources – to enter these distorted markets. Legislators and regulators are facing the challenge to define and implement transparency and resulting market opening.

To achieve the objective of creating and enforcing functioning energy markets and eventually a single European internal energy market three consecutive legislative packages were adopted in the European Union between 1996 and 2009. The process which led to the three packages will be described and analysed in this chapter. As of today, packages two and three have not yet been fully transposed into national legislation in a significant number of Member States and with significant impact on market functioning. And it seems quite likely that stronger enforcement of existing legislation will be necessary, and probably further strengthening of legislation.

The common aim was to establish the internal energy market by removing obstacles and barriers for new market entrants, aligning norms, standards and regulations and facilitating consumers' choice between competing providers. The objective is ensuring market functioning with fair access and consumer protection, where incumbent oligopolies have never known real competition. On 4 February 2011, the European Council of Ministers – once more – emphasized the need for swift completion of the internal energy market – now to be achieved by 2014 (Council Conclusions 2011a). As it stands today, timely achievement cannot be taken for granted.

7.2   THE FIRST INTERNAL ENERGY MARKET PACKAGE

Fifteen years ago, after several years of discussions in and among EU Member States, in the European Parliament and among stakeholders representing various parts of the energy sector, a first legislative package was adopted, consisting of two directives, the Electricity Market Directive 1996 and the Gas Market Directive 1998. The package codified agreement about basic principles of the internal energy market, but it fell short of concrete and effective regulations. Nearly no measures are included, which could really be enforced.

The Electricity Market Directive 1996 (and the Gas Market Directive 1998 in a similar way) reiterated the need *"for smooth running of the internal market"* (Recital 2), stated that *"the completion of a competitive electricity market is an important step towards completion of the internal market"* (Recital 3), but it also included some precautionary reservations. *"The internal market in electricity needs to be established gradually, in order to enable the industry to adjust in a flexible and ordered manner to its new environment and to take account of the different ways in which the electricity systems are organized at present"* (Recital 5). It must favour *"interconnection and interoperability of systems"* (Recital 6). Reference is made to basic aims of the European Union, such as cohesion (Recital 20) and the subsidiarity principle (Recital 11).

Progress towards market functioning is meant to be achieved, when the need for *"common rules for the production of electricity and the operation of electricity transmission and distribution systems"* is introduced (Recital 22), as well as the necessity to take into account *"the position of autoproducers and independent producers"* (Recital 24), and the requirement that *"the transmission system operator must behave in an objective, transparent and non-discriminatory manner"* (Recital 25). This aspect was detailed in Article 20, in normative language and not setting up clear regulations: *"Member States shall ensure that the parties negotiate in good faith and that none of them abuses its negotiating position by preventing successful outcome of negotiations"* (Article 20, 2).

The directive introduces forward looking principles which are still accepted today, which could be beneficial for the growth of renewable energy, such as the possibility to give priority *"to the production of electricity from renewable sources"* (Recital 28 and Article 8, 3). And it states that *"any abuse of a dominant position or any predatory behaviour should be avoided"* (Recital 37). The legislators were, however, aware that *"some obstacles to trade in electricity between Member States will nevertheless remain in place* (Recital 39) so that the European Commission is asked to report about potential improvements.

The directive defines *"General rules for the organization of the sector"* (Chapter II), and it describes how authorization procedures for new generation capacity should be organised in a transparent and competitive way, either using an *"authorization procedure"* (Article 5) or a *"tendering procedure"* (Article 6).

In Chapter IV *"Transmission system operation"* the independence of system operators from electricity producers is required. Member States are obliged to *"designate [. . .] a system operator to be responsible for operating, ensuring the maintenance of, and, if necessary developing the transmission system in a given area and its interconnectors with other systems, in order to guarantee security of supply* (Article 7, 1). The system operator is required to provide sufficient information, and *"[. . .] shall not discriminate between system users [. . .] particularly in favour of its subsidiaries or shareholders"* (Article 7, 5), and *"[. . .] shall be independent at least in management terms from other activities not relating to the transmission system"* (Article 7, 6).

The directive establishes the principle of *"unbundling"*, which continues to be a key challenge for market functioning all over the European energy sector, meaning the separation of production and transmission by defining principles which should be applied by Member States by taking *"the necessary steps to ensure"* (Article 14) at least the separation of accounts. *"Integrated electricity undertakings shall, in their internal accounting, keep separate accounts for their generation, transmission and distribution activities [. . .]"* (Article 14, 3). Member States shall *"ensure that there is no flow of information between the single buyer activities of vertically integrated electricity undertakings and their generation and distribution activities"* (Article 15, 2). It soon became evident that the unbundling regulations were among the weakest parts of the directive, although they constituted the maximum that Council and European Parliament could agree on. This weakness was one of the major reasons for continued discussions and for changes of these regulations in the second and third packages.

The Electricity Market Directive 1996 had to be transposed into national law not later than 19 February 1999, with additional time of one or two years for three Member States. The Gas Market Directive 1998 had to be transposed by 11 August 2000. Several critical regulations were overhauled in the second and third package, before they were fully transposed by Member States.

7.3 THE SECOND INTERNAL ENERGY MARKET PACKAGE

Soon after the first package had entered into force, discussions started about necessary improvements to accelerate internal market implementation. In March 2001, the European Commission presented a first proposal for amendments, which were discussed in the European Parliament and in the Council. Several changes and amendments were proposed, until the European Commission presented a revised draft in June 2002. Political agreement about major content of the second package was reached among the Member States' energy ministers on 25 November 2002. Remaining details – like dates for mandatory market opening – were decided in February 2003. Eventually, the European Parliament agreed and the new package entered into force on 4 August 2003.

The package consists of the Electricity Market Directive 2003, the Gas Market Directive 2003 and Regulation 1228/2003. Main changes and improvements compared to the first package are concrete dates for gas and electricity market opening for industrial and for private consumers, improvements of the authorization procedures for small and distributed generation connected to the electricity distribution system, specifications regarding independent system operators, ownership unbundling, and labelling requirements for electricity suppliers.

7.3.1 Trying to overcome the weaknesses of the first package

The Electricity Market Directive 2003 (and similarly the Gas Market Directive 2003) includes – in the recitals – a number of analytical points to be considered and improved, if the weaknesses of the first package are to be cured. It is stated that *"important shortcomings and possibilities for improving the functioning of the market remain, notably concrete provisions are needed to ensure a level playing field in generation and to reduce the risks of market dominance and predatory behaviour, ensuring non-discriminatory transmission and distribution tariffs, through access to the network on the basis of tariffs published prior to their entry into force, and ensuring that the rights of small and vulnerable customers are protected and that information on energy sources for electricity generation is disclosed, as well as reference to sources, where available, giving information on their environmental impact"* (Recital 2). Referring to decisions of Council and European Parliament important findings are quoted. *"The main obstacles in arriving at a fully operational and competitive internal market relate amongst other things to issues of access to the network, tarification issues and different degrees of market opening between Member States"* (Recital 5). And more precisely: *"For competition to function, network access must be non-discriminatory, transparent and fairly priced"* (Recital 6).

An important issue that is addressed but not fully solved (Recital 8) is the independence of the system operator. It is acknowledged that it *"is appropriate that the distribution and transmission systems are operated through legally separate entities where vertically integrated undertakings exist"* and it *"is necessary that the independence of the distribution system operators and the transmission system operators be guaranteed especially with regard to generation and supply interests"*. However, the directive does not implement the most obvious conclusion from these statements – due to continued opposition from important Member States. The directive simply states the obvious. *"It is important however to distinguish between such legal separation and ownership unbundling. Legal separation does not imply a change of ownership of assets and nothing prevents similar or identical employment conditions applying throughout the whole of the vertically integrated undertakings"*. Given these limitations, only an appeal to Member States can be issued to try the impossible: *"However, a non-discriminatory decision-making process should be ensured through organisational measures regarding the independence of the decision-makers responsible"*.

In addition to these main findings, the need for effective regulation is highlighted. *"It is important that the regulatory authorities in all Member States share the same minimum set of competences"* (Recital 15). The European Commission's intention to set up a *"European Regulators Group for Electricity and Gas"* (ERGEG) is welcomed (Recital 16) and the need for ensuring that *"customers should be able to choose their supplier freely"* (Recital 20) is reiterated. As a result of these deliberations the following articles were agreed for the directive.

7.3.2 Consumer protection

In addition to what had already been outlined in the first package, Article 3 now specifies the possibility for Member States to impose public service obligations to undertakings of the electricity[1] sector, which have to be *"clearly defined, transparent, non discriminatory, verifiable and*

---

[1] In this chapter most of the specific regulations are referred to as they apply for the electricity sector. As the problems of market opening, unbundling, transparency, consumer protection etc. are basically the same as in the gas sector, the gas sector is not explicitly (and by way of repeating regulations for the electricity sector) described and analysed.

shall guarantee equality of access for EU electricity companies to national consumers" (Article 3, 2). *"Member States shall ensure that all household customers, and, where Member States deem it appropriate, small enterprises, (namely enterprises with fewer than 50 occupied persons and an annual turnover or balance sheet not exceeding EUR 10 million), enjoy universal service, that is the right to be supplied with electricity of a specified quality within their territory at reasonable, easily and clearly comparable and transparent prices"* (Article 3, 3). Member States are obliged to ensure *"adequate safeguards to protect vulnerable customers"* (Article 3, 5 and Annex A).

In order to enable consumers to take informed decisions the directive requires reliable labelling of the electricity sold to consumers. Suppliers must disclose in the electricity bills *"the contribution of each energy source to the overall fuel mix of the supplier over the preceding year"* (Article 3, 6(a)) or *"at least the reference to exiting reference sources, such as web-pages, where information on the environmental impact, in terms of at least emissions of CO_2 and the radioactive waste resulting from the electricity produced by the overall fuel mix of the supplier over the preceding year is publicly available"* (Article 3, 6(b)).

7.3.3 *Transmission System Operators (TSOs)*

The tasks of transmission system operators are defined in Article 9. They are responsible for *"ensuring the long-term ability of the system"*, for *"contributing to security of supply through adequate transmission capacity and system reliability"*. TSOs have to manage *"energy flows on the system"*, including taking into account of interconnections and assuming responsibility for maintaining the system and providing ancillary services. And they have to ensure *"non-discrimination [. . .] particularly in favour of its related undertakings"*. And they are responsible for *"providing system users with the information they need for efficient access to the system"*.

It is well known that transparency and independence have not been the main assets of vertically integrated utilities, part of which operated the transmission system. Now they are obliged to non-discrimination against competitors. This is why the Electricity Market Directive specifies a set of criteria for independence and minimum requirements to be implemented by Member States. Article 10 includes a list of criteria which is longer than in Electricity Market Directive 1996. *"Where the transmission system operator is part of a vertically integrated undertaking, it shall be independent at least in terms of its legal form, organisation and decision making from other activities not relating to transmission"*. But a reservation remains: *"These rules shall not create an obligation to separate the ownership of assets of the transmission system from the vertically integrated undertaking"*. In order to compensate for the lack of ownership separation, minimum criteria for *"legal unbundling"* (as this form of partial separation of ownership and decision making is called) are set up in order to *"ensure the independence of the transmission system operation"*. These criteria comprise non-identity of decision makers of TSO and producing units of the company, capability of acting independently and regular reports. But again, there are limitations: *"This should not prevent the existence of appropriate coordination mechanisms to ensure that the economic and management supervision rights of the parent company in respect of return on assets [. . .] in a subsidiary are protected."* The parent company shall even have the right to *"approve the annual financial plan"*, but *"It shall not permit the parent company to give instructions regarding day-to-day operations, nor with respect to individual decisions concerning the construction or upgrading of transmission lines"*. It is obvious that the latter enables the mother company to influence decisions of the TSO about large-scale investment, which may become necessary to operate and extend the grid in a sense of non-discrimination.

It should be mentioned that the regulations about *"Unbundling of accounts"* (Article 19) as already established in the 1996 directive, are basically maintained.

7.3.4 *Market opening*

Whereas the possibility for Member States to *"require the system operator, when dispatching generating installations, to give priority to generating installations using renewable energy sources or waste or producing combined heat & and power"* is reasonably maintained in the directive as it was in 1996, another aspect regrettably also remains unchanged: *"A Member State may, for reasons of security of supply, direct that priority be given to the dispatch of generating installations using indigenous primary energy fuel sources, to an extent not exceeding in any calendar year 15% of the overall primary energy necessary to produce the electricity consumed in the Member State concerned".* This was – when it was enacted in 1996 – a regulation to secure a minimum use of coal and oblige system operators to comply with this objective. And it has not been changed since then, although it is at least no longer necessary, if not a hurdle against renewable energy and other independent production.

An important step towards a real internal market is marked by the possibility for electricity and gas suppliers to enter Member States' markets and for customers to freely choose their suppliers. The 2003 directive is very specific about *"Third party access"* to the grid system (Article 20). Access must be *"based on published tariffs, applicable to all eligible customers and applied objectively and without discrimination between system users. Member States shall ensure that these tariffs, or the methodologies underlying their calculation, are approved prior to their entry into force in accordance with Article 23 and that these tariffs, and the methodologies — where only methodologies are approved — are published prior to their entry into force"* (Article 20, 1). And the 2003-directives grant new electricity and gas suppliers the right to enter markets in Member States. And they set the dates for costumers to freely choose their suppliers. For *"all non-household customers"* this applies from 1 July 2004 (Article 21, 1b), for domestic consumers the date is 1 July 2007 (Article 21, 1c).

Market Opening is an important step forward. It is not completely implemented, however. Article 26 of the Electricity Market Directive 2003 and Article 27 of the Gas Market Directive 2003 foresee some *"Derogations"*, setting up a formal procedure for which derogation may be taken into account and how the permission process will be. *"Member States which can demonstrate, after the Directive has been brought into force, that there are substantial problems for the operation of their small isolated systems, may apply for derogations from the relevant provisions [. . .], as far as refurbishing, upgrading and expansion of existing capacity are concerned, which may be granted to them by the Commission. The latter shall inform the Member States of those applications prior to taking a decision, taking into account respect for confidentiality. This decision shall be published in the Official Journal of the European Union. This Article shall also be applicable to Luxembourg"* (Article 26, 1).

7.4   MOVING TO THE NEXT PACKAGE

As foreseen in the 2003 directives, the European Commission presented a report on the implementation of the directives (European Commission 2005a) before 1 January 2006. The report is very clear on what has been achieved and where Member States fail to comply. *"With the adoption of the second Electricity and Gas Directives the basic framework for the development of a real internal market is in place. It is now for Member States to implement the directives effectively and make the market work in practice. This needs to be done in the spirit of the objective of the directives, i.e. to create a functioning internal energy market in the interest of EU citizens".* The report sums up: *"This report sets out the reasons why this objective is still far from being reached"* (page 15). The Commission therefore announces to present two more in depth reports by the end of 2006 and early 2007 (European Commission 2007a and European Commission 2007b).

The 2005 report states that *"the EU now has the unique opportunity to create the largest integrated competitive electricity and gas market in the world. The EU cannot afford to miss the opportunity of making this market a success".* One year after the directives should have been fully transposed into Member States' law and one year and a half before *"markets will be fully open to competition"*, major drawbacks can be observed. *"The most important persisting shortcoming is the lack of integration between national markets. Key indicators in this respect are the absence of price convergence across the EU and the low level of cross-border trade. This is generally due to the existence of barriers to entry, inadequate use of existing infrastructure and – in the case of electricity – insufficient interconnection between many Member States, leading to congestion. Moreover, many national markets display a high degree of concentration of the industry, impeding the development of effective competition".* The report finds that *"most Member States have transposed the new Directives only with delay and some not yet at all"* (pages 2–3). Little progress is seen, in particular no efforts exceeding the minimum requirements of the directive are observed, with some exception only in the *"Nordpool"* area. After most Member States have missed the deadline, *"a number of the structural measures will come into effect later than provided in the directives. Particularly important in this respect are the rules on regulatory oversight and the unbundling provisions"* (page 4). The Commission foresees to report in 2007 whether additional legislation is deemed necessary, *"for example additional unbundling or further powers to Regulators"* (page 3).

The Commission sees insufficient progress in market integration, with low level of cross-border trade, high price differences of partly more than 100% between Member States, with insufficient interconnection capacities being a major reason for the delays. The report recalls that *"back in 2002 the European Council in Barcelona adopted the objective that all Member States must have interconnection capacity equivalent to at least 10% of their national consumption, an objective which has not yet been achieved"* (page 5). Planning procedures for interconnection capacities are lagging far behind, and non-discriminatory use of existing infrastructure is not yet achieved. Despite unbundling regulations, *"capacity reservations in favour of historical long-term contracts"* (page 6) add to the problem and are seen as incompatible with non-discrimination.

Industry remains basically national, cross border trade and competition of electricity and gas is rather an exception. Exposure to EU-wide competition has not been achieved in most markets. Concentration is high and even further consolidating since market opening started. And it is still difficult for new market players to enter the market and *"only a very limited share of new electricity generation projects were commissioned by nonincumbents"* (page 6). Where cross-border acquisitions take place, this is *"aggravating the risks associated with concentration"* (page 8), particularly through merging of incumbent companies.

A particularly problematic area of non-implementation is the independence of network operators. Although *"effective unbundling of network operation from the competitive parts of the business is essential to ensure independent network operation and non-discriminatory access to networks for all market participants"* (page 11), this has widely not been fully implemented. On the other hand, for the TSOs about half of the Member States and a few in the gas market have gone beyond the minimum requirements of the directive and have moved to ownership unbundling. Table 7.1 gives an overview of the implementation status in the EU Member States for the independence of network operators for the gas and electricity sector. For the distribution networks most Member States have made use of derogations granted by the directive or have at least not gone beyond the minimum requirements. And network access for new market participants remains costly and resulting in many complaints.

The Commission pressed hard for implementation of the energy market directives. Two more reports on progress and prospects for the internal electricity and gas markets were presented in early 2007 (European Commission 2007a, European Commission 2007b). The main messages of these two communications were also included in a strategic communication *"An Energy Policy for Europe"* (European Commission 2007c), which was published in parallel.

Summing up the results of a market inquiry in the Member States (European Commission 2007b) the Commission finds that *"the objectives of market opening have not yet been achieved"* (page 2). *"The shortcomings identified in these key areas call for urgent action and priority should be given to four areas: (1) achieving effective unbundling of network and supply activities, (2) removing the regulatory gaps (in particular for cross border issues), (3) addressing market concentration and barriers to entry, and (4) increasing transparency in market operations"* (page 3). Details of the enquiry, which was launched on 17 June 2005, followed by preliminary report on 16 February 2006 and a public consultation based thereon are given in the report.

Table 7.1.   Status of legal and ownership unbundling.*

Member State	Electricity transmission	Gas transmission
Austria	legal	legal
Belgium	legal	legal
Denmark	ownership	ownership
Finland	ownership	
France	legal	legal
Germany	legal	partly legal
Greece	legal	
Ireland	legal	unbundling not implemented
Italy	ownership	legal
Luxembourg	legal	unbundling not implemented
Netherlands	ownership	ownership
Portugal	legal	
Spain	ownership	legal
Sweden	ownership	ownership
UK	ownership	ownership
Norway	ownership	
Estonia	legal	unbundling not implemented
Latvia	legal	unbundling not implemented
Lithuania	ownership	unbundling not implemented
Poland	legal	unbundling not implemented
Czech Republic	ownership	unbundling not implemented
Slovakia	legal	unbundling not implemented
Hungary	ownership	legal
Slovenia	ownership	unbundling not implemented
Cyprus		
Malta		

*Data taken from European Commission 2005a, p. 12 – the sections without specification have derogation from the unbundling provisions.

And it certainly did not come as a surprise that *"generally speaking the vertically integrated incumbent companies were not in favour of further measures, whilst consumers, traders/new entrants and authorities supported the call for legislative initiatives"* (page 5). The inquiry finds little progress regarding national scope of electricity markets and high levels of concentration of the pre-liberalization period. Customers' choice is still very limited; there remains a *"dependency on vertically integrated incumbents for services throughout the supply chain"* (page 5). Market integration is hampered by insufficient interconnector capacities in combination with capacity reservations. The report also complains about *"information asymmetry between the vertically integrated incumbents and their competitors"* (page 7).

There are specific shortcomings in balancing markets. *"Currently, balancing markets often favour incumbents and create obstacles for newcomers. The size of the current balancing zones is too small, which leads to increased costs and protects the market power of incumbents"* (page 8).

Remedies are seen in enlarging the geographical size of control areas. Furthermore the report suggests full enforcement of the directives and of competition law. Well-designed anti-trust regulations would help to limit the *"market power of pre-liberalisation monopolies"* (page 9). *"Market partitioning remains one of the most serious obstacles to market integration. The fight against collusion between incumbents remains a priority of antitrust enforcement action, reflecting the overall priority of the Commission to fight attempts by undertakings to coordinate rather than to compete"* (page 11).

Other findings are related to still insufficient levels of unbundling and the need to strengthen regulators. This should also result in more transparent market information, which again would reduce the dominance of the incumbents.

Derived from the market inquiry, the Communication on *"Prospects for the internal gas and electricity market"* (European Commission 2007a) draws several conclusion on how to overcome the *"Improper implementation of the current legal framework"* (page 6) and how to further improve the legal framework. And it is made clear that infringement procedures will be pursued for non-implementation. According to European Commission 2007a, 34 infringements against 20 Member States were launched for non-transposition or violation of the existing directives. And 26 more procedures against 16 Member States including all the biggest were launched by sending – as a first step – reasoned opinions to be answered in due time.

The *"main deficiencies observed in transposition of the new internal market directives"* (page 6) read like a complete list of what should have been done but – for reasons to be discussed later – was not implemented in many Member States. These deficiencies are:

- *"Regulated prices preventing entry from new market players;*
- *Insufficient unbundling of transmission and distribution system operators which cannot guarantee their independence;*
- *Discriminatory third party access to the network, in particular as regards preferential access being granted to incumbents for historical long term contracts:*
- *Insufficient competences of the regulators;*
- *No information given to the Commission on public service obligations, especially as regards regulated supply tariffs:*
- *Insufficient indication of the origin of electricity, which is essential in particular for the promotion of renewable energy"* (page 6).

The conclusions drawn by the Commission are very clear and therefore quoted here in total: *"The persistent nature of these infringements, almost two years and a half after the obligation to transpose the directives on 1 July 2004, clearly demonstrates the insufficiencies and shortcomings of the current EC legal framework arising from the directives. Energy regulators are not granted the necessary powers and independence enabling them to ensure that open markets that function in an efficient and non discriminatory manner are put into place. In addition, the existing legal framework does not allow for a proper and efficient regulation of the cross border issues relating to gas and electricity network access. The preferential access that is granted in a persisting manner to cross border interconnectors clearly demonstrates the shortcoming of the current rules. Finally, the legal and functional unbundling of network operators that are vertically integrated with production and supply activities, which is provided for under the current directives, does not succeed in ensuring equal access to the networks for all suppliers"* (page 6–7).

The report is particularly outspoken, when it comes to unbundling, which is a main point for all those Member States which do not plan to significantly limit the influence of the incumbent utilities. The Commission indicates that evidence shows that present legislation – although the problems are not universal and it has led to some progress – is not sufficient. *"Inherently, legal unbundling does not suppress the conflict of interest that stems from vertical integration, with the risk that networks are seen as strategic assets serving the commercial interest of the integrated entity, not the overall interest of network customers"* (page 10). The Commission therefore underlines *"that only strong unbundling provisions would be able to provide the right incentives for system operators to operate and develop the network in the interest of all users"* (page 11). Two main avenues are considered for further TSO unbundling.

The first avenue is *"fully (ownership) unbundled TSOs"*, the second is *"separate system operators without ownership unbundling"* (page 11), i.e. the independent system operator (ISO). A number of advantages are seen in the first alternative, such independence would be inherent in this model with non-discrimination for third party access being *"guaranteed and perceived as such"* and *"investment decisions would no longer be distorted by supply interests"*. Fully unbundled TSOs would more easily tend to cross-border cooperation or merging. And there is a direct relation between the level of necessary regulations and the degree of unbundling: *"Full ownership unbundling would reduce the need for increasingly burdensome regulation as the regulatory oversight could be less detailed to ensure that no discrimination takes place"* (page 11).

The ISO solution would require formal separation of system operation from ownership of the assets. But this solution *"would require detailed regulation and permanent regulatory monitoring"* (page 11).

The Commission is looking for a solution that can be applied in all Member States. The fact that – at the time of the report – already 11 Member States had introduced ownership unbundling is considered to be a strong argument in favour of this solution, even more so, because *"economic evidence shows that ownership unbundling is the most effective means to ensure choice for energy users and encourage investment"* (page 12). And the *"independent system operator approach would improve the status quo but would require more detailed, prescriptive and costly regulation and would be less effective in addressing the disincentives to invest in networks"* (page 12).

Another major finding of the report is the need for more coordination of system operators. It is therefore considered to strengthen existing networks of TSOs and of Gas Transmission Europe to achieve progress about standards, frameworks and eventually move towards evolution of regional system operators.

In September 2007, the European Commission officially published proposals for amending the Electricity and Gas Market directives of 2003, accompanied by a detailed explanatory memorandum (Electricity Market Directive 2007, Gas Market Directive 2007). Starting from the overall assessment that *"the process of developing real competitive markets is far from complete"* (page 2) the memorandum outlines the major challenges to tackled. Regarding unbundling they could build on strong support by the European Parliament considering that *"transmission ownership unbundling is the most effective tool to promote investments in infrastructures in a non-discriminatory way, fair access to grid for new entrants and transparency in the market"* (European Parliament 2007, page 4), and by the Council of European Energy Regulators (CEER)[2].

The proposals of the European Commission focussed on six major aspects, which were outlined in detail in the Explanatory Memorandum. Draft texts for amending the directives were included (European Commission 2007a):

- Effective separation of supply and production activities from network operations (pages 4–7);
- Enhanced powers and independence of national regulators (pages 7–9);
- An independent mechanism for national regulators to cooperate and take decisions: the Agency for the Cooperation of Energy Regulators (pages 9–13);
- Efficient Cooperation between Transmission System Operators (pages 13–15);
- Improving the functioning of the market (pages 15–19);
- Cooperation to reinforce security of supply (pages 19–20);

Like in earlier years, the first and foremost challenge to be addressed is the separation of production and networks. And the Commission knows that there are some Member States which will not easily give in. The Memorandum holds that there is a *"fundamental conflict of interest within integrated companies"* so that a *"transmission system operator may treat its affiliated companies better than competing third parties"*. Consequently, *"non-discriminatory access to information cannot be guaranteed"* (page 4). Integrated TSOs have *"an inherent interest to limit new investments when this will benefit its competitors"* (page 4). The Commission concludes that *"more effective unbundling of transmission system operators is therefore clearly necessary"* underlining *"that the preferred option of the Commission remains ownership unbundling"*. The Commission nevertheless considers and outlines *"an alternative option for Member States that*

---

[2]CEER published on 6 June 2007 a set of six papers taking position on the main issues in the new energy legislation (CEER 2007a–CEER 2007f).

choose not to go down this path" (page 5). The approach is the *"Independent System Operator"* (ISO), an option which allows integrated companies to retain ownership of the network and at the same time act *"truly independently"* as a network operator. This option indispensably requires that *"regulation and permanent regulatory monitoring must be put in place"* (page 6). The options apply for both electricity and gas sectors, although the Commission takes note of the fact that unbundling is more advanced in the electricity sector. Finally, it needs to be said that these unbundling requirements must be applicable all over the European Union, for companies based in the EU and for companies from third countries (planning to be) active in the EU. It is underlined that *"no supply or production company active anywhere in the EU can own or operate a transmission system in any Member State of the EU"* (page 7).

To enforce market liberalisation and fair access for market players it is of paramount importance that regulatory authorities are established in all Member States, *"with substantial powers and resources, allowing them to ensure proper market regulation"*. The Commission suggests strengthening the regulators' *"market regulation powers"* (page 8). In addition these authorities need a clear mandate for cooperation at European level, particularly for cooperation with ACER, the Agency for the Cooperation of European Energy Regulators[3]. This agency should be created resulting from the good experience in the past with ERGEG, the European Regulators Group for Electricity and Gas.

The proposals (Electricity Market Directive 2007 and Gas Market Directive 2007) were discussed controversially – particularly regarding unbundling and ISO – in the European Parliament and in the Council. It took until April 2009 to agree on the third – and so far last – internal energy market package.

7.5   THE THIRD INTERNAL ENERGY MARKET PACKAGE

The Package consists of the amended directives for the internal electricity and gas market (Electricity Market Directive 2009 and Gas Market Directive 2009), a Regulation establishing ACER (Regulation 713/2009) and two regulations on conditions for access to the gas and electricity grids (Regulation 714/2009 and Regulation 715/2009), establishing i.a. the cooperation between the various network operators via ENTSO-E and ENTSO-G. The new package entered into force on 3 March 2011. In September 2011, the European Commission started 17 infringement procedures for not fully transposing the Electricity Market Directive and 18 for not complying with the Gas Market Directive.

7.5.1   *Electricity and Gas Market directives overhauled*

The amended directives, agreed on and published in 2009, intended to considerably move forward the establishment of an open and liberalised internal energy market with a level playing field for all market players in all Member States. Both directives strongly emphasize the need for achieving a well-functioning internal market. They underline the necessity to remove remaining obstacles for market opening: *"In particular, non-discriminatory network access and an equally effective level of regulatory supervision in each Member State do not yet exist"* (Electricity Market Directive 2009, Recital 4). The directive includes the reasoning from the Explanatory Memorandum of Electricity Market Directive 2007 that the *"inherent risk of discrimination"* (Recital 9) needs to be removed, and calls for *"effective unbundling"* (Recital 9), particularly in *"removing any conflict of interests between producers, suppliers and transmission system operators"* (Recital 12). Although the Commission and many stakeholders had advocated for ownership unbundling as

---

[3]The Agency www.acer.europa.eu was established – following the proposal outlined in European Commission 2007a – as the EU's independent body for the completion of the internal energy market, both for electricity and gas. It was officially launched in March 2011. The headquarters of the Agency is in Ljubljana, Slovenia, established after ERGEG was officially dissolved by Electricity Market Directive 2009.

the least onerous and most efficient solution, the alternative of establishing an Independent System Operator was kept in the directive. Accordingly, *"Member States should therefore be given a choice between ownership unbundling and setting up a system operator or transmission operator which is independent from supply and generation interests"* (Recital 17). However, doubts remain. Deliberations are included, enumerating various preconditions for the ISO really being independent to *"be ensured by way of specific additional rules"* (Recital 19). Detailed regulations about securing either ownership unbundling or independence of ISO are provided in Articles 9–16 of the directive.

Independence of the TSOs is not the only necessary element of a level playing field. Member States should therefore encourage decentralised generation and energy efficiency. Furthermore: *"Authorisation procedures should not lead to an administrative burden disproportionate to the size and potential impact of electricity producers. Unduly lengthy authorisation procedures may constitute a barrier to access for new market entrants"* (Recital 31). And: *"Further measures should be taken in order to ensure transparent and non-discriminatory tariffs for access to networks. Those tariffs should be applicable to all system users on a non-discriminatory basis"* (Recital 32).

For achieving all this independent regulators are indispensable. Articles 35–40 contain detailed lists of objectives for the independent regulators, their decision process and about record keeping. *"Each Member State shall designate a single national regulatory authority at national level"* (Article 35, 1). The regulator should be able to fix and approve tariffs and underlying methodologies for network access and use. *"In carrying out those tasks, national regulatory authorities should ensure that transmission and distribution tariffs are non-discriminatory and cost-reflective, and should take account of the long-term, marginal, avoided network costs from distributed generation and demand-side management measures"* (Recital 36). Such tariffs are essential for effective market functioning, particularly for new and independent players. The regulators are entitled to take reasonable measures for *"facilitating access to the network for new generation capacity, in particular removing barriers that could prevent access for new market entrants and of electricity from renewable energy sources"* (Article 36, paragraph 1e). The regulator should therefore be empowered to *"impose effective, proportionate and dissuasive penalties"* (Recital 37), and should have *"the power to request relevant information from electricity undertakings, make appropriate and sufficient investigations and settle disputes"* (Recital 38). ACER shall have a significant role in coordinating and advising Member States' regulators in this regard.

The final provisions of the directive allow Member States to apply for derogations (Article 44, 1) from some of the provisions of the directive. Some of the unbundling provisions do not apply for the island states Cyprus and Malta and for Luxembourg (Article 44, 2). Finally, as it is obvious that full effectiveness and application of the directive cannot be taken for granted the directive contains reporting and review clauses. The Commission shall report to Council and Parliament by 3 March 2013 the latest *"outlining the extent to which the unbundling requirements under Chapter V have been successful in ensuring full and effective independence of transmission system operators, using effective and efficient unbundling as a benchmark"* (Article 47, 3). More interestingly, by 3 March 2014, the next revision of the directives might start. *"Where appropriate, and in particular in the event that the detailed specific report referred to in paragraph 3 determines that the conditions referred to in paragraph 4 have not been guaranteed in practice, the Commission shall submit proposals to the European Parliament and the Council to ensure fully effective independence of transmission system operators by 3 March 2014"* (Article 47, 5).

7.5.2 Establishing ACER

ACER was established as an independent body for regulation on national and on European level. It is an independent legal personality with administrative and financial autonomy. The task is defined in Regulation 713/2009: *"The Agency should monitor regional cooperation between transmission system operators in the electricity and gas sectors as well as the execution of the tasks of the European Network of Transmission System Operators for Electricity (ENTSO for*

Electricity), and the European Network of Transmission System Operators for Gas (ENTSO for Gas). The involvement of the Agency is essential in order to ensure that the cooperation between transmission system operators proceeds in an efficient and transparent way for the benefit of the internal markets in electricity and natural gas" (Recital 7). *"The Agency has an important role in developing framework guidelines which are non-binding by nature (framework guidelines) with which network codes must be in line. It is also considered appropriate for the Agency, and consistent with its purpose, to have a role in reviewing network codes (both when created and upon modification) to ensure that they are in line with the framework guidelines, before it may recommend them to the Commission for adoption"* (Recital 9).

Independence of the Agency is of paramount importance. *"The Agency should have the necessary powers to perform its regulatory functions in an efficient, transparent, reasoned and, above all, independent manner. The independence of the Agency from electricity and gas producers and transmission and distribution system operators is not only a key principle of good governance but also a fundamental condition to ensure market confidence"* (Recital 18). Further details are in Articles 4–11.

The Agency was officially launched in March 2011 with the headquarters located in the Slovenian Capital Ljubljana. Since then, it has become a major actor in designing and implementing the internal energy market.

7.6   STILL TO BE ACHIEVED: COMPLETING THE INTERNAL MARKET

The implementation of an internal energy market remains high on the agenda. It is a frequent topic in energy working groups, Council meetings and among stakeholders. Legal transposition of the 2009 directives is lagging behind. According to a "Non-paper" (European Commission 2011), there are still important barriers in place, such as highly concentrated markets, regulated household prices and in particular, unbundling of the networks has not sufficiently progressed. On 1 February 2011, a month before the directives had to be fully transposed in all Member States, not a single one had notified transposition to the Commission. *"Only in a few Member States had draft legislation been submitted to Parliament for adoption, or had the government been empowered by the Parliament to adopt the necessary transposition measures, but the measures themselves had not been adopted as yet. There is a clear risk that there will be delays in the transposition. More in general, the state of implementation of internal market legislation at national level is overall disappointing, with currently over 60 infringement proceedings underway on the second internal energy package alone"* (page 3–4). This devastating interim assessment provides the groundwork for ongoing discussions.

7.6.1   *Some progress until 2012*

Regularly, the European Council comes back to the pending issues regarding the internal energy market, assessing anew what needs to be done and outlining new time-tables. On 4 February 2011, the Heads of State and Government agreed on strong conclusions (Council 2011a). *"The EU needs a fully functioning, interconnected and integrated internal energy market. Legislation on the internal energy market must therefore be speedily and fully implemented by Member States in full respect of the agreed deadlines"* (page 1). Underlining the urgency the Council set a new deadline for full implementation: *"The internal market should be completed by 2014 so as to allow gas and electricity to flow freely."* Furthermore: *"No EU Member State should remain isolated from the European gas and electricity networks after 2015 or see its energy security jeopardized by lack of the appropriate connections"* (page 2).

As if the Heads of State and Government were becoming impatient, they came back to the lack of implementation of the directives despite urgent decisions to be taken about the financial and monetary crisis. At their spring Council, on 1–2 March 2012, they highlighted some key points, one of them being the energy market completion. *"As regards energy, it is important to implement the guidelines agreed in February and December 2011, delivering on the commitment to complete the internal energy market by 2014, including through the full implementation of the Third Energy Package in recognition of agreed deadlines, and to interconnect networks across borders. The European Council looks forward to the Commission's communication expected by next June assessing the degree of liberalisation and integration of the internal energy market".* (Council 2012a, page 7).

In October 2012, the European Commission listed 12 key actions to make the single market work. *"Key Action 4"* is to *"Improve the implementation and enforcement of the third energy package and make cross-border markets that benefit consumers a reality"* (European Commission 2012l, page 8–9). Further detailing the related challenges, on 15 November 2012, the European Commission presented a new report about the status of implementation of the internal market directives for electricity and gas (European Commission 2012m). The Communication, titled *"Making the internal energy market work"*, reiterates the benefits of integrated European energy markets. It underlines that *"by 2014 the existing legislation needs to be implemented fully"* and that *"today the EU is not on track to meet this deadline"* (page 2). According to one of the accompanying Staff Working Document (European Commission 2012o), the energy markets are still largely concentrated. In eight Member States, the historic incumbents still control most of the power generation – more than more than 80% in eight Member States and more than 50% in another three (Table 7.2).

The Communication (European Commission 2012m, pages 3–6) describes some progress achieved (more consumer choice, more transparency and market coupling, some cross-border

Table 7.2.   Structure of the electricity market 2010.*

Member State	Number of companies producing at least 5% of the net generation	Market share (%) of the largest generator in the electricity market
Cyprus	1	100
Malta	1	100
Estonia	1	89
Latvia	1	88
France	1	86.5
Luxembourg	2	85.4
Greece	1	85.1
Slovakia	1	80.9
Belgium	3	79.1
Czech Republic	1	73
Slovenia	2	56.3
Portugal	2	47.2
Denmark	2	46
Hungary	3	42.1
Sweden	5	42
Romania	6	35.6
Lithuania	5	35.4
Ireland	6	34
Germany	4	28.4
Italy	5	28
Finland	4	26.6
Spain	4	24
United Kingdom	9	20
Poland	5	17.4
Bulgaria	5	???
Netherlands	5	???
Austria	4	???

*Data are taken from European Commission 2012o, Table 10, page 42.

price convergence, common regulatory best practises and technical standards), but sees too little progress when it comes to competition and access to transmission grids. *"The lack of open and non-discriminatory access to transmission infrastructure has prevented new entrants from competing fairly in the market"* (page 6). Although a new industry branch with *"transmission-only focus and increasingly cross-border footprint"* is developing, a lot more needs to be achieved. So far only a minority of Member States have started certification of their TSOs as either ISO or ownership unbundled. This is what the Commission calls the *"enforcement challenge"* (page 7), which needs to be *"tackled urgently to complete the internal market by 2014"*. The Commission is actively pursuing Member States by infringement procedures for not (fully) transposing the second and third energy market package into national law.

Table 7.3 shows how many cases were still pending at the end of 2012. As of 29 October 2012, seven Member States still had infringement cases pending for not fully transposing Electricity Market Directive 2003 and seven for the Gas Market Directive 2003, four of them (Ireland, Greece, Poland, UK) regarding both directives. In addition, 13 Member States had infringement cases pending for non-transposition of the Electricity Market Directive 2009 and for the Gas Market Directive 2009. Full transposition still is to be enforced and the European Commission needs to provide regular updates about progress achieved. The Energy Ministers' Council of 3 December 2012 (Council 2012b) endorse the Commission's intention and welcomes the announced measures for facilitating and enforcing the directives. A more in depth discussion was scheduled for

Table 7.3.   Infringement cases for non-transposition of Packages Two and Three.*

	2nd Energy Market Package		3rd Energy Market Package	
	Electricity	Gas	Electricity	Gas
Belgium	Cases closed	Cases closed	Case closed	Case closed
Bulgaria	Cases closed	1 case pending	Case pending	Case pending
Czech Republic	Cases closed	Cases closed	No case	No case
Denmark	Cases closed	Cases closed	Case closed	Case closed
Germany	1 case pending	Cases closed	No case	No case
Estonia	Cases closed	Cases closed	Case pending	Case pending
Ireland	1 case pending	1 case pending	Case pending	Case pending
Greece	1 case pending	1 case pending	No case	No case
Spain	Cases closed	Cases closed	Case closed	Case closed
France	Cases closed	1 case pending	Case closed	Case closed
Italy	1 case pending	Cases closed	No case	No case
Cyprus	No case	No case	Case pending	Case pending
Latvia	Cases closed	Cases closed	No case	No case
Lithuania	Cases closed	Cases closed	Case pending	Case pending
Luxembourg	Cases closed	Cases closed	Case pending	Case pending
Hungary	Cases closed	Cases closed	No case	No case
Malta	Cases closed	No case	No case	No case
Netherlands	Cases closed	Cases closed	Case closed	Case closed
Austria	Cases closed	Cases closed	Case closed	Case closed
Poland	1 case pending	2 cases pending	Case pending	Case pending
Portugal	Cases closed	Cases closed	No case	No case
Romania	Cases closed	1 case pending	Case pending	Case pending
Slovenia	Cases closed	Cases closed	Case pending	Case pending
Slovakia	Cases closed	Cases closed	Case pending	Case pending
Finland	Cases closed	Cases closed	Case pending	Case pending
Sweden	1 case pending	Cases closed	Case pending	Case pending
United Kingdom	1 case pending	1 case pending	Case pending	Case pending

*Data are from European Commission 2012o, page 155.

22 February 2013, foreseen to prepare the European Council in May 2013 and to be adopted by the Energy Council on 7 June 2013[4].

7.6.2   Chicken or egg: renewable energies in the internal market

The internal energy market is not an aim as such. Most stakeholders and politicians agree on this statement. The idea of the single internal European energy market is driven by the objective of a transparent energy sector, open to new entrants, with fair competition and resulting cost reductions and benefits for consumers. In such an open and fair market, costs would basically be borne by those who use the energy and the polluter-pays-principle would apply, meaning that externalities would be fully internalized in energy prices. Greenhouse gas emissions and resulting climate change would be part of the energy prices, e.g. by taxation and/or by a functioning emissions trading system which would lead to relevant and effectively incentivising carbon prices. Market mechanisms would be in place, which fully take into account the qualities of clean and sustainable energy. Energy savings and efficiency would be facilitated by a price structure responsive to demand changes and incentivising reduced consumption.

Markets functioning in this regard can provide a level playing field, resulting in renewable energy becoming more and more and fossil and nuclear sources less competitive due to increased carbon costs and included societal costs for health damages, waste procurement and other aspects, which are borne by the general public through tax payments in traditional energy markets. Markets, where incumbents are dominant, rules and costs not transparent and balancing areas and regulations following the logic of fossil and nuclear energy, are certainly not appropriate to send and receive price signals for running a system based on renewable energy – particularly variable wind and solar. This is why discussions are ongoing about how energy markets must be designed so that they can accommodate high shares of flexible power generation and provide the necessary signals for flexibility options, storage, and grid infrastructure – generally speaking for running energy systems based on clean and sustainable renewable energy.

These necessary changes have not yet been addressed in depth in the debates about completing the internal energy market by 2014, and even less have they been considered by many of those who continue to insist that renewable energy has to be exposed to market risks and should no longer be hedged by support mechanisms. However, it is obvious that structural market distortions, in and between Member States prevail so that renewable energy needs support, although an increasing number of technologies is becoming cost competitive at an increasing number of sites.

In most parts of Europe and beyond, wind power onshore is the most mature and most cost competitive among the renewable energy technologies. Wind industry is keen to participate in markets, if the markets allow for fair competition – and even with some distortions remaining. In a paper published in 2012, the European Wind Energy Association (EWEA) has elaborated on necessary measures to accelerate the completion of the single market and to make the markets viable for wind power and other renewable sources (EWEA 2012). Whilst striving for completing the internal market by 2014, as foreseen by the energy market directives, EWEA finds that *"structural market distortions remain the main obstacle to creating an internal energy market and integrating wind energy [. . .] Integration of large amounts of wind energy in a cost efficient manner requires changing the current market arrangements"*. EWEA underlines that the Renewables Directive 2009 is of high importance for development of wind and other renewables. The supportive provisions are needed *"due to a number of market and regulatory failures or imperfections. Thus, support mechanisms for wind and other renewables should be seen in the context of an unfinished liberalisation and as compensation for numerous market failures that arise from an internal market that is fragmented, dysfunctional and far from fully developed. Prevailing market distortions – in the forms of continued massive subsidies to fossil fuels and*

---

[4]Some of the discussions will only take place after this book is finalized, and potential decisions, which will most likely underline the need for completing the internal market by 2014 and meeting all foreseen deadlines, will also be taken then.

nuclear energy, market concentration and regulated prices – together with market rules that do not consider wind energy characteristics create increased market risks for wind energy generators".

Another important aspect is addressed in EWEA's analysis. Whereas the European Union has agreed to strive for climate protection, greenhouse gas reduction and future oriented competitive and clean and renewable energy, legislation and market rules are not following suite. Decisions to support renewables are taken, legislation enacted, but support for fossil and nuclear continues with increasing amounts of subsidies – much higher than for renewables. According to the International Energy Agency (IEA), in 2012 more than 380 billion Euros (500 billion USD) were spent on subsidies for fossil fuels compared to 60 billion Euros (80 billion USD) for renewable energy[5].

As a logical consequence, EWEA underlines that *"full exposure of wind energy generators to market risks can only take place under the preconditions that markets are functional, competitive, liquid and transparent and that all technologies are exposed to the same conditions on a level playing field".* This certainly requires removal of fossil fuel and nuclear power subsidies, but market functioning also needs to be adapted to completion – in reality and for all market players, not only for renewable energy producers. *"A functional, mature and competitive market should be seen as a pre-condition to exposing wind generators and other producers to market risks, included carbon and fuel price risks"* (EWEA 2012, page 5–6). A functional and mature energy market would certainly be a market without fossil and nuclear subsidies, without intransparent cost calculations and grid structure. Instead, it would encourage new participants. Larger shares of energy would be traded at commercial power exchanges, with larger control zones in interconnected markets, shorter trading time horizons with functional intraday and balancing markets. Only in such flexible and short term markets could the value and quality of variable and flexible sources be properly assessed and become a significant element of cost and price calculations. In such markets, wind, solar and other abundant variable renewable sources would be used whenever available. This is desirable because wind and solar energy has no resource costs and thus reduces prices at stock exchanges via the merit order effect. Other energy sources with higher marginal costs should be used according to their availability and flexibility for providing system stability – together with storage capacities of different size and reaction times and grid enforcement for tapping the benefits of larger balancing areas.

7.6.3   Outlook

Completing the internal energy market – even if not by 2014 – will be a major step forward towards a level playing field enabling new market entrants and particularly renewable energy producers to fairly compete for market shares in a renewables based energy system. Without removing market dominance of incumbent oligopolies new technologies and new actors will continue to need enabling frameworks to facilitate their market entrance. Removal of subsidies for unsustainable energy sources and designing grids, storage capacities and markets according to the requirements of energy systems with very high shares of renewables are further important steps towards a clean and sustainable energy supply. Section II of this book will elaborate on scenarios and debates related to this objective.

---

[5]The IEA-figures are only on fossil fuel subsidies, nuclear subsidies are not included. The fact of six times higher subsidies for fossil fuels then for renewables – which has been described by IEA since 2009 http://www.worldenergyoutlook.org/resources/energysubsidies/ is even more striking on this basis. The figures quoted here are from a presentation given by IEA's chief economist, Fatih Birol, at the World Future Energy Summit in January 2013, in Abu Dhabi/United Arab Emirates.

Section II
The way forward: 2020 and beyond

CHAPTER 8

Scenario development and policy debates

Rainer Hinrichs-Rahlwes

The 2020 framework was put in place after years of debates in the European Union and among Member States. Although the March 2007 Summit of Europe's heads of state and government certainly played a pivotal role in establishing the consensus and later transforming it into relevant legislation in the form of the Climate and Energy Package (Chapter 3 of this book), it is worthwhile understanding that this level of agreement – including both painful compromise and joyful success for some of the participants – would not have been possible without prior scientific work on the feasibility of the targeted framework and on potential impacts on costs, jobs, environmental and other issues. The climate and energy package was well prepared by various stakeholders and it became reality through decisive and courageous action of decision makers in governments and parliaments all over Europe, underpinned by the expertise of EU officials and of scientists looking more deeply into impacts and solutions.

This section explores the importance of focussed policy debates as well as the important role of learning from the experience of other parts of the world and of sharing experience that might help improve performance towards a sustainable energy system elsewhere. And it bridges the gap from today's state of implementation of the directives via a number of scenarios to the ongoing policy debates which are to result in agreement about a post-2020 framework for renewable energy, energy efficiency and greenhouse gas emissions reduction in the European Union.

Chapter 9 addresses the role of scenarios for successfully reaching agreement on the climate and energy package in 2009. Scenarios looking beyond 2020 aiming at 2050 to outline a vision of a sustainable energy system are described and analysed. It is show how assumptions, methodology but also the intention of the authors can influence the results of a study. It is shown how Eurelectric in 2009, supported later in a separate study by EWI, re-opened the debate on a number of questions which had been settled in the climate and energy package – immediately after the package was agreed. The agreement on strengthening national support systems for renewables and supporting convergence by good practice exchange was attacked by forging proof for billions of Euros of potential savings through harmonisation of support schemes, and upholding the role of nuclear and CCS as allegedly necessary for climate protection and security of supply and for reducing costs. Eurelectric's *Power Choices Study* (Eurelectric 2009) and EWI's *European RES-E Policy Analysis* (EWI 2010) are confronted with reactions from other stakeholders and compared to studies aiming at greenhouse gas reductions of 80–95% by 2050 while at the same time increasing the shares of renewable energy – including scenarios aiming at 100% renewables in 2050.

Apart from stakeholder contributions the three energy related roadmaps presented by European Commission in 2011 are described and analysed. The *Low Carbon Roadmap* (European Commission 2011a), the *Transport Roadmap* (European Commission 2011c1) and the *Energy Roadmap 2050* (European Commission 2011d) provide a solid framework for assessing policy options for a new climate and energy package beyond 2020. The roadmaps are evaluated in detail, in particular their strengths and weaknesses are analysed and solutions provided.

In **Chapter 10, Christine Lins,** Executive Secretary of the global policy network, REN21, provides a broader view on the European developments from a global perspective. She shows how renewable energy development has become a global ambition with significant benefits for sustainable development and climate change mitigation. She argues that the initiative of the UN Secretary General *(Sustainable Energy for All – SE4ALL)* aiming at doubling the global share of renewables until 2030 is comparable to the early EU process of setting indicative targets as described in Chapter 2 of this book). The debate about new and binding 2030-targets for the EU and SE4ALL might as well be mutually supportive, as both are setting a mid-term milestone, which is important to add to long term visions.

Eventually, **Chapter 11** describes ongoing discussion about how to keep the European Union on track towards a fully sustainable energy system by 2050 the latest. Recent Communications and legislative proposals from the European Commission are analysed, such as the RES-Strategy (European Commission 2012b), the ETS-Backloading Proposal (European Commission 2012i), the ILUC amendment proposal (European Commission 2012f), the State Aid Modernisation Process (e.g. European Commission 2012a, European Commission 2013), to envisaged guidance papers on the reform of renewables support schemes and on the cooperation mechanisms, the Green Paper on creating a post-2020 climate and energy framework (European Commission 2013f) and the ongoing debate, fuelled particularly by Renewable Energy Industry (EREC 2013b), some Environmental NGOs, and by some MEPs about the need for a new and binding 2030 Renewables target. Related debates in Council and Parliament will be referred to. This chapter prepares the programmatic outlook developed in the final part of the book, which provides a view on what needs to be done to keep Europe on track towards a fully sustainable energy system, fuelled only by renewables. Recommendations are provided on which policy decisions will be needed to move forward towards 100% renewable energy – by 2050 the latest.

CHAPTER 9

Scenarios up to 2050 – assumptions, figures and more

Rainer Hinrichs-Rahlwes

9.1 INTRODUCTION: SCENARIO OVERVIEW

In public debates, energy scenarios, like other scenarios, are sometimes mistaken for forecasts of what will inevitably happen. And they are often used by decision makers to describe and explain (and facilitate) what should happen. They are used to understand the logic and variables of past and future events and how development could be influenced. Scenarios provide information on how to facilitate desirable developments by changing parameters. Scenarios are developed to understand underlying processes, and they are refined to better assess potential levers to shift development in one or the other direction. In addition, scenarios often include variables regarding impact of development on and from other areas, e.g. the impact of certain energy shares on economic growth, jobs, greenhouse gas emissions or supply security or the sensitivity of renewables growth to GDP development and vice versa.

Scenarios are sometimes presented to the public without highlighting – or even without mentioning – the underlying assumptions, without specifying which figures are given inputs and which are results of the scenario calculations. For assessing the quality and content of scenarios and their potential value for informed policy decisions, it is necessary to be aware of these factors, to know the underlying research (or policy) questions and the assumptions about technical and economic potential of certain sources and applications. Assumptions and extrapolations about technology learning curves, regional distribution, availability of resources and technology, quality and capacity of infrastructure and many other aspects can significantly influence the output of seemingly pure mathematical iterations. It is important to know what kind of differentiation is taken into account by a scenario and which elements are neglected and/or considered on a levelized basis.

Scenario calculations can be useful for making informed decisions, if input and output factors and the underlying interests and objectives are transparent. Scenarios can be supportive for outlining how certain targets can be achieved and which impact this could have on specific factors. In particular, if scenarios are used for defining – material or political – costs of potential decisions, it is of paramount importance to define a reference case, which is a major challenge to the real value of a scenario. For example, some scenarios have provided up to three-digit billion figures for the amount of Euros to be spent for renewable energy development and deployment. If these figures are used without calculating benefits of renewable energy e.g. avoided fuel imports, avoided carbon emissions, reduced health costs, job creation, or at least the investment that would be needed to replace out-dated existing fossil and nuclear power production by new conventional power plants, they will most likely result in severe doubts about investment in renewable energy. If, however, these aspects and other impacts are included, macro-economic benefits are nearly always evident – and micro-economic and personal benefits are most likely foreseeable for the near future or they are also evident already.

There are various global energy scenarios, 164 of them are described and analysed in IPCC 2012. Most of these scenarios which are calculating global development of different energy technologies and greenhouse gas emissions see renewable energy expanding – at widely different levels though – even in reference scenarios. All scenarios, particularly those aiming at significant greenhouse gas reduction, result in widespread and considerable growth of renewable energy until 2050, which is the most frequent timeline for long-term outlook. Scenarios generally indicate a widespread distribution of renewable energy around the world, although there is a broad variation regarding the spread over different countries and regions. The scenarios also have a wide range of technology diversification, with different results regarding dominating technologies and the overall capacities and shares of renewable energy deployed. The global scenarios vary widely regarding the expected penetration level of renewable energy in the different sectors, ranging – for the global overall energy consumption – from 23% in 2030 and not significantly more in 2050 in some baseline scenarios (e.g. IEA-WEO 2009) to 48% in 2030 and more than 90% in 2050 in ambitious greenhouse gas reduction scenarios (Energy [R]evolution 2010).

For the purpose of this book, the global scenarios shall only serve as a background to emphasize that there is a wide range of scenarios being developed and discussed globally, with methodology and underlying assumptions being acknowledged and further refined by those who develop scenarios for Europe.

Several scenarios will be described and analysed in some detail, which were used – or even developed – for intervention in the debate about the EU climate and energy package and the 2020-targets or for developing the next steps beyond 2020, particularly a vision for 2050 and – more recently – for providing input to the debate about a new milestone up to 2030 for sustainable European climate and energy policies.

This chapter describes and evaluates scenarios meeting one of the following categories, where knowledge about their fitting into one of them helps assessing strengths and weaknesses:

- Scenarios developed to underpin the decisions about the 2020-targets are designed around the guiding question how (not if) the targets – particularly the 20% RE-target and the greenhouse gas reduction targets – can be achieved and how efforts can be shared between the 27 Member States.
- Scenarios looking towards 2050 are often calculated with the underlying assumption and/or objective that a greenhouse gas reduction of *"80 to 95%"* compared to 1990 levels must be achieved (except for some reference scenarios).
- Some scenarios are set up to demonstrate how 100% renewable energy can be reached by 2050 and which benefits this would provide.

9.2   SCENARIOS FOR 2020: PREPARING AND ASSESSING THE 2020-TARGET

When the climate and energy package of 2008 was developed, the targets were not just invented by visionary politicians, but the range was discussed and elaborated beforehand by scientific analysis, including scenarios estimating technical and economical potentials of different renewable energy sources for different application in different EU Member States. The same applies for costs of various pathways and their potential benefits such as greenhouse gas reduction, job creation and reducing import dependency. Discussions and iterations developed within the framework of an EU-supported project *"futures-e"* had a major impact on the development and particularly the acceptance of a 20%-target for 2020. They provided major groundwork for evaluating and assessing the feasibility of the 20% target based on technology and cost assumptions. Such iterations were also used for splitting the joint target among the Member States alongside objective criteria like resource distribution and economic potential, which was labelled as *"effort sharing"* and as such became part of the Renewables Directive of 2009. The respective scenarios are described in Futures-e 2008 and Futures-e 2009.

For the purpose of this chapter, which is looking beyond the existing policy framework and targets for 2020, the results as described in Futures-e 2008 and more in depth in Futures-e 2009 should provide a rough guidance about the deliberations which eventually led to the policy framework and targets agreed by the European Union (Chapter 3 of this book). And it will be demonstrated that – due to limited experience at that time – some assumptions later turned out to clearly underestimate cost increases of conventional energy sources as well as cost reductions and resulting higher potentials for some more innovative renewable energy sources.

Table 9.1.   Fossil energy price assumptions (in USD$_{2005}$/barrel of oil equivalent).

	2005*	2010**	2010***	2015**	2015***	2020**	2020***
Oil	54	44.59	54.5	44.95	57.9	48.08	61.1
Gas	30.31	33.86	41.5	34.22	43.4	36.99	46.0
Coal	13.32	12.53	13.7	13.38	14.3	14.1	14.7

*Reference (Futures-e 2008, page 8); **Futures-e 2008 (PRIMES); ***2009 (PRIMES).

The objective of Futures-e 2008 is highlighted by the authors: *"This study presents a balanced scenario to meet Europe's renewable energy commitment. It provides an assessment on the effects of a 20% RES target in terms of final energy demand in the year 2020 within the European Union (EU27)"* (page 1). Based on an updated *"Green-X balanced scenario"*[1] as develop by the European Commission in the Renewable Energy Roadmap, published in January 2007 (European Commission 2007) the calculations were meant to comply with recent policy decisions and therefore be useful for the decision process. Wherever possible, data were not newly generated but updated, based on the same methodology which was used for the roadmap. As a result the scenarios are comparable with the roadmap, which makes them easy to apply for policy decisions. The research aimed at *"identification of the technology-portfolio of a 20% RES target for the sectors electricity, heat and transport – meeting criteria such as cost-effectiveness and future perspectives"* (page 2). In addition, the costs of reaching a 20% target should be calculated as well as the benefits of avoided fossil fuel use. Avoided CO_2 emissions were calculated as well as country specific RES deployment.

The scenario was developed to assess the impact of a policy decision in favour of a 20% target for 2020, which was assumed to be very ambitious for the EU and particularly for some Member States. Therefore, the modelling was done assuming *"that all Member States apply immediately (i.e. from 2008 on) efficient & effective support policies for RES, setting incentives on technology level, accompanied by strong energy efficiency measures to reduce the overall growth of energy demand"* (page 4). In order not to underestimate costs *"negative additional costs appearing on technology level by country are not counted"* (page 4). The finally presented main scenario evaluates a broad range of technology options and policy settings. In order to be compatible with other Commission calculations, some input figures were taken from the PRIMES model: sectoral energy demand, primary energy prices, conventional portfolio and efficiencies, CO_2-intensity of sectors. The data were matched with those specifically defined for the study: 20% target, reference electricity prices, RES costs, RES potential, biomass import restrictions, technology diffusion and learning rates (page 5).

Fossil fuel and reference energy prices were taken from the PRIMES Scenario of 2006. They were much lower than real energy prices in 2008, and from today's perspective projections for 2010, 2015 and 2020 (see Table 9.1) are incredibly low. This was modified later in Futures-e 2009, which further developed this study.

Futures-e 2009 additionally included a *"high price case"* with oil prices going up to USD 76.4 in 2010, USD 88.1 in 2015 and USD 100 in 2020. It is obvious that the development of fossil fuel prices is a sensitive factor for calculating deployment rates and incremental costs of renewable energy sources. As part of fossil fuel prices, carbon costs are another relevant element. While in reality they were floating between 7 and 30 €/t, CO_2-prices were set at 20 €/t in Futures-e 2008. Following the PRIMES scenarios, the assumption is a price of 20 € in 2005 and 2010 in Futures-e 2009, increasing to € 26.3 (in 2015) and € 34.5 in 2020. Given today's development of prices below 5 €/t, even close to zero, due to abundant certificate availability on the markets, higher

---

[1]The Green-X model was developed in the framework of an EC-funded project. For details see http://www.green-x.at/

carbon prices might be considered a very optimistic assumption, which would nevertheless need to come true in order to revitalise the emissions trading system.

Another important aspect of realistic target setting is the potential of renewable energy, in the European Union and in each Member State. Futures-e 2008 therefore evaluates a *"realisable mid-term potential"* (page 10), which is available *"assuming that all existing barriers can be overcome and all driving forces are active"*. Special attention is paid to the availability of biomass, which plays a major role in all three sectors. The most significant generation of renewable electricity until 2020 (page 35–36) is seen in the big Member States France (173.8 TWh/a), Germany (195 TWh/a), Spain (136.4 TWh/a) and United Kingdom (134.5 TWh/a), with Italy (118.2 TWh/a) and Sweden (102.6 TWh/a) following. Austria, Finland, Poland, Romania, the Netherlands and Portugal are also estimated to have a potential of more than 30 TWh/a. Hydropower is expected to contribute 384 TWh, followed by wind onshore (256.5 TWh), solid biomass (217 TWh) and wind offshore (195 TWh). Photovoltaics are estimated at 26 TWh.

For heating and cooling the largest yield is expected in Germany (195 TWh), France (174 TWh), Spain (136 TWh) and UK (135 TWh), followed by Italy (118 TWh) and Sweden (102 TWh). Finland and Italy are also expected to produce more than 80 TWh. Solid biomass, off-grid and grid connected, is envisaged to contribute more than 950 TWh, leaving 220 TWh to all other sources with solar (90 TWh) the strongest contributor.

For the transport sector, a share of 9.9% of gasoline and diesel demand is expected to be achieved in 2020, very close to the 10% target of the Renewables Directive for renewables in transport.

The expected overall final energy consumption expected in 2020 is approaching the 20% overall target and the 10% biofuels target, which the European Commission had proposed to be part of the directive. Table 9.2 shows, where Futures-e 2008 expected higher shares than the proposed targets and where the targets would not be achieved under this scenario. These differences were a main element of later agreement on the targets, when higher economic potential – as in most of the Western European countries – was considered to be a reason to allow for a target above the scenario results and relatively low economic potential – as in most of the Eastern European Member States – was taken as a reason to set the target below the scenario outcomes[2]. And it was a starting point for the cooperation mechanisms of the directive.

The scenarios presented in Futures-e 2008 (and Futures-e 2009) also provide cost estimations for reaching the proposed targets. This is calculated in some detail, based on so-called cost bands, not on average costs per technology. Based on this approach the range of marginal costs for renewable energy technologies and sources in all three sectors is calculated and presented in the study. As a result, a range of costs for each technology is presented (page 14). In the heating and cooling sector all technologies and sources – under certain conditions – are seen to be cost competitive with the average market prices, but more expensive where these conditions do not apply. None of the biofuel options is considered to be cost competitive with the average diesel or gasoline prices. In the electricity sector, some technologies – under some conditions – are close to or below the average market price already (wind onshore, hydro, biowaste, biogas and biomass cofiring). On the other hand, all of them figure under much higher costs in other conditions. Wind offshore and tide and wave are estimated to have significantly higher costs. The costs of photovoltaic electricity production are calculated at a range from 1,430 to 1,640 €/MWh, far beyond the next expensive technology (solar thermal) – an assessment which from today's perspective shows how learning curves for PV were underestimated only a few years ago.

Based on these cost ranges the scenario iterations result in a specific technology breakdown per country and for the EU as a whole. Again, this provides interesting insight how much the expected development and cost decreases for some technologies were overestimated and some

---

[2]This approach does not explain the comparatively very low targets for Sweden and Finland, or the targets for Denmark and Austria. Eventually – as explained in Chapter 3 of this book – all targets were a result of the negotiations between governments and parliament.

Table 9.2.   Proposed 2020-targets and scenario results for 2020*.

Country	Proposed target [%]	Scenario result [%]	Scenario/target (percentage points)
Austria	34	35.8	+1.8
Belgium	13	9.3	−3.7
Denmark	30	34.4	+4.4
Finland	38	46.3	+8.3
France	23	23.8	+0.3
Germany	18	16.7	−1.3
Greece	18	19.4	+1.4
Ireland	16	15.6	−0.4
Italy	17	14.2	−2.8
Luxembourg	11	7.4	−3.6
Netherlands	14	10.5	−3.5
Portugal	31	33.3	+2.3
Spain	20	21.1	+1.1
Sweden	49	58.8	+9.8
United Kingdom	15	13.9	−1.1
Cyprus	13	12.7.	−0.3
Czech Republic	13	13.9	+0.9
Estonia	25	27.7	+2.7
Hungary	13	13.6	+0.6
Latvia	42	38.3	−3.7
Lithuania	23	25.1	+2.1
Malta	10	11.1	+1.1
Poland	15	17	+2.0
Slovakia	14	15.7	+2.7
Slovenia	25	29.5	+4.5
Bulgaria	16	18.8	+2.8
Romania	24	25	+1.0

*Figures for targets and scenario results from Futures-e 2008, pages 35–36.

others clearly underestimated in 2008, compared to today's reality (Futures-e 2008, page 21). The scenario calculates average annual growth rate (AAGR) for the recent past and for the immediate future (2006–2010) and for the two five-year periods from 2010–2015 and from 2015–2020. The figures are further developed and treated under a range of different assumptions on national and/or harmonised support mechanisms. There are some gradual deviations in Futures-e 2009 regarding the following figures from 2011. The basic tendency, however, of e.g. assuming strong growth rates for offshore wind and lower rates for onshore, seeing advantages of solar thermal compared to photovoltaic power production remains and for offshore wind is even more optimistic than Futures-e 2008.

In the electricity sector, the highest growth rates for the whole period (2005–2020) are expected for two large-scale and centralised technologies. Solar thermal electricity with 33.2% AAGR and offshore windpower 32.4%, followed by tidal and wave power with 26.8% and PV with 19.2%. At the lower end, there is hydropower (with 0.9% for large scale and 2.2% for small scale installations) and geothermal power. The most spectacular aberration from today's reality is the expectation of wind offshore growth, set at more than 43% from 2006–2010, 37.7% from 2010–2015, and 19.5% until 2020, whereas onshore wind is expected to grow at reduced rates of only 2.9% from 2015 to 2020, down from 14.4% from 2006 to 2010. The scenario expects hydropower (large and small) still to provide the largest share of renewable electricity in 2020 (31%, but down from 51% in 2010), closely followed by wind onshore with 21% (like in 2010), solid biomass 18%, up from 17%) and wind offshore (16%, up from only 2% in 2010). PV, like solar thermal electricity and tide and wave remain at 1% of the renewable power in 2020, up from nearly nothing in 2010. From today's perspective it seems that growth rates and deployment for offshore wind will stay significantly behind these expectations. Cost decreases are much slower than expected. On the other hand, onshore wind keeps steadily growing, in contrast to slow development of solar thermal electricity and huge increase of photovoltaic electricity production, which can still be expected to further decrease costs.

For the heating and cooling sector, the expected changes are not as big as in the electricity sector. And the deviations from real developments are less spectacular than in the power sector. Various forms of biomass are expected to remain the dominant sources for renewable heat, with non-grid solid biomass contributing 64% in 2020 (down from 74% in 2010) followed by solid biomass for district heating (grid-connected) with 17% in 2020 (up from 14% in 2010). Solar thermal heating and solar thermal heat pumps are expected to perform at high growth rates (17.5% for solar and 9.7% for heat pumps) over the whole period of 15 years, but their share remains small, with only 8% for solar thermal in 2020 (up from 2% in 2010) and 3% for heat pumps (up from 1%).

In the transport sector, the scenario considers biofuels only, not taking into account electric cars and other innovative forms of renewables in transport. Given today's reality the approach may seem to be a bit conservative, but on the other hand, electric cars are still far from a breakthrough. Within the biofuels sector, however, there is a strong deviation from today's reality. The scenarios expect *"advanced biofuels"* to grow by an average of 34.6% annually until 2020, thus providing 55% of a 10% share of biofuels in 2020 (up from 11% of a 1% share in 2010). Imports are expected to provide 34% (2010) and 29% of the biofuels in transport. From today's perspective, the development of advanced (or second generation) biofuels is clearly lagging behind these expectations. And at the same time, traditional (*"first generation"*) biofuels are attacked for not being sustainable – by oil-companies and environmental NGOs, which results in reduced public acceptance and thus reduced support.

The scenario also calculates avoided greenhouse gas emissions from new installed renewables capacity from 2006 to 2020 (page 28f.), estimating emissions reductions of 209 million tonnes in 2010 and 756 million tonnes in 2020, which equals 14% of the EU's total greenhouse gas emissions of 1990. Together with existing RES installations the avoidance would amount to 1,403 million tonnes of greenhouse gas emissions, equalling 25% of the EU's 1990 emissions.

Another positive impact of renewables is avoiding fossil fuel imports. The study estimates cumulated savings of € 287 billion from 2006 to 2020 (page 28), increasing from € 1.6 billion in 2006 (equalling 0.02% of Europe GDP) to € 39 billion (equalling 0.28% of GDP). Investment of € 538 billion will be more or less evenly spread over the period (page 29), the largest share of it (63%) expected to go into RES-E, whereas 21% is the forecast for heating and cooling, 11% for CHP and 5% for RES-T. In contrast, additional generation costs for meeting the 2020-target of 20% are calculated to be € 10.9 billion (page 31), which is supposed to be much lower in reality, due to the very low cost assumptions for conventional energy on which these calculations are based. Furthermore, it is expected that the electricity prices will not significantly rise, because a *"significant part of additional generation costs and costs for grid extension and system operations would be recovered by the reduction of wholesale electricity prices obtained from increased RES-E generation"* (page 32).

Futures-e 2009 provides further details and some modifications, which basically confirm the results of Futures-e 2008. In an important aspect, however, this study goes beyond the approach of the 2008 study. Potential developments are calculated using different scenarios regarding the degree of policy ambition and harmonisation of support mechanisms in Europe. Apart from a business-as-usual scenario (assuming continued national RES-support), three scenarios are examined assuming harmonisation of support systems (a technology neutral quota system, a technology specific quota system and a technology specific feed-in premium system), as well as an option called *"strengthened national RES support by 2011"* (pages 53ff). All scenarios are evaluated regarding RES deployment, cost development, technology spread and efficiency for target reaching. Major variances are seen in offshore wind development and photovoltaics, both of which grow much stronger in strengthened national support scenario. As a result of various iterations, the *"strengthened national RES support by 2011"* is the scientists' preferred option, resulting in most stable growth and relatively low costs.

This assessment was refined later – and is also being further developed, including more specific options regarding feed-in and quota systems, in a new EU-supported project *"Beyond 2020 – the future of support systems"*[3], which is partly coordinated by some of the scientist who produced the studies Futures-e 2008 and Futures-e 2009. The new project is supposed to provide significant input to the ongoing discussions about allegedly necessary harmonisation vs. optimisation or phase-out of renewables support mechanisms in Europe.

9.3 2030: EVALUATING AND QUESTIONING THE 2020 FRAMEWORK

The agreement between European Parliament and Council about the Renewables Directive seemed have settled the question of whether or not a harmonised European support mechanism for renewable energy in the electricity sector should be introduced. For the review foreseen in 2014, the directive foresees that the functioning of national support schemes and the 2020 target must not be questioned by potential proposals from the European Commission meant to improve implementation. The agreement, however, did not stop the discussion among scientists and stakeholders. Even before the directive was fully transposed into national law and even before the first experience was available for evaluation, the Institute of Energy Economics at the University of Cologne (EWI) published a report claiming to provide scientific evidence that a harmonised quota system (HQS) would deliver on the 2020-target much more efficiently and targeted than the existing policy mix (EWI 2010). The report was welcomed and used in the political debate by those who had long since advocated harmonisation of support for renewables, based on a specific understanding of market functioning, which would – at the end of the day – deliver least cost results. Some of them frankly added that only with quota systems growth rates could be effectively controlled and limited. By those, who were striving to accelerate renewables deployment and cost decreases by fine-tuning and further developing policy frameworks, the study was criticised for unrealistic assumptions and therefore misleading results (as an example see Futures-e 2010).

9.3.1 *Forging proof for harmonisation gains*

EWI 2010 claims to analyse *"regional RES-E deployment induced by different support schemes and the interaction between the renewable and the conventional power markets under growing RES-E shares until 2020, with a further outlook until 2030"* for the EU27 and including Norway and Switzerland, which are not Members of the EU but participating in the European power market. The authors create a Business-as-usual scenario and a related *"Linear Optimization Model for Renewable Electricity Integration in Europe (LORELEI)"* (page 10). They intend to model – partly by using auxiliary scenarios – the different approaches to renewables support:

- price-based versus quantity based,
- technology-specific versus technology neutral, and
- national versus EU-wide harmonised support.

To take into account *"a certain bandwidth of design options"* the study *"analyzed in detail"* three scenarios (pages 10–11):

- a harmonised quota system (HQS),
- a business-as-usual (BAU) scenario, which models existing schemes as they are actually designed, and

---

[3]The project website http://www.res-policy-beyond2020.eu/ provides further overview about interim results and progress of the project.

- a *"Cluster scenario"*, which combines a joint quota system in those Member States which use quota systems today, leaving those with other mechanisms like in the BAU-scenario.

Based on these assumptions and settings and on related calculations the expected result is presented as a *"conclusion"*: *"Since the HQS is the only scenario which reaches the European RES-E targets due to the quantity-based setting, and since it does so at minimal RES-E generation cost, this normative scenario is at the center of our discussion"* (page 11). This *"conclusion"* takes a setting (the given quota) as a result. And it neglects – at least – ongoing developments and adaptations of existing support schemes (including potential use of cooperation mechanisms) as well as an extreme unlikeliness of the *"Cluster scenario"* in reality.

There are other elements in the study which show the biased approach. Repeatedly, the study mentions *"intermitting"* wind and solar power – not taking into account that most scientific literature has meanwhile turned to the more appropriate term *"variable"* renewable energies, because wind and sun availability can – increasingly well – be predicted and thus become part of dispatch planning. It is therefore neither *"intermittent"*[4] nor *"intermitting"* like an unpredictable failure or a blackout. Based on the misperception of wind and solar as highly unreliable the authors foresee *"a hardly reduced demand for conventional power capacity in order to fulfill the required security of supply. Altogether the total installed (renewable and conventional) generating capacity rises significantly to fulfill both the RES-E targets and the system adequacy criterion"* (page 14).

The study finds as a result of the HQS-scenario, what is long since known elsewhere and which is a major reason for applying technology specific support mechanisms. They find that *"still expensive technologies like photovoltaics or geothermal power are hardly deployed under technology-neutral support"* (page 12), underpinning this by wind onshore as the dominant technology (42%), followed by wind offshore (21%), which is modelled so that it is not among the *"still expensive technologies"*, and biomass (24%). Among the *"less mature technologies"* concentrating solar power (6%) has the highest share; PV is only marginal – again based on cost-learning-curve assumptions for these technologies, which have not been observed in reality. As a result of these calculations, in the HQS-scenario *"mainly countries with high wind potential deploy RES-E above their national target"* and *"altogether the 12 Eastern European countries are net exporters of TGCs due to their comparatively low RES-E targets"*. Reality tends to provide just the opposite as has been shown in Chapter 4. Overshooting national targets seems much more likely, where well-designed technology specific support frameworks are in place.

This does not prevent the authors from further detailing their approach. *"[. . .] harmonization gains are defined as cost savings in RES-E generation [. . .] solely through a switch from national to harmonised support"*, but they have to admit that these gains *"may be counteracted by additional costs of e.g. grid enhancements due to higher concentration of intermitting RES-E in certain regions resulting from harmonized support. Grid costs and additional costs in the conventional power systems are not considered in this study, but would need to be assessed to find an overall efficient solution"* (page 13). In fact, neglecting these additional costs is a major weakness of the study, which – properly assessed – might easily turn around the results from gains to significant losses.

With high shares of wind (and solar) *"the utilization of the installed conventional power capacities is reduced"* and a *"shift towards a higher share of peak load capacity"* is expected (pages 12–13).

The main message of the study is eventually provided by comparing the BAU-scenario with the other (more) harmonised options which were described earlier. Not surprisingly, the authors find *"that HQS contains potential efficiency gains"* (page 15). To present their 174 billion €$_{2007}$ *"total RES-E cost savings of a switch from BAU"*, arising *"from the change from a national to a EU-harmonized support"* and from *"the change from a mainly technology-specific to a*

---

[4]This is the English word used by other studies conducted in the anglophonic world.

technology-neutral support of RES-E in Europe" they even create an *"auxiliary HQS scenario"* to make assumptions comparable – but unfortunately not more realistic (pages 15–16). Consequently, they find that even their Cluster scenario would still provide cost savings of 35 billion €$_{2007}$ (page 16). Eventually, it goes without saying that – with some methodological reservations in order not to forget elementary scientific standards – they conclude what should be proven by the study: *"A more harmonized approach would enable to utilize considerable cost-savings in RES-E generation, as a result of competition between plant locations. In addition, the introduction of competition between technologies would lead to substantial cost-savings"* (page 16).

9.3.2 *Harmonisation gains in question*

As EWI 2010 was obviously developed to provide new input to a debate which had been politically settled when the Renewables Directive was agreed on, their study could not remain uncontested. Gustav Resch and Mario Ragwitz, authors of several studies for the European Commission on scenario development and support scheme assessment provided an answer which was widely agreed on in the policy debate about the future of support schemes. In the framework of the REshaping project[5] they published – as a direct reply to EWI 2010 – a paper titled *"Quo(ta) vadis, Europe?"* (REShaping 2010) Summarizing their findings they show *"that the efficiency gains through a move from national RE support schemes to an EU wide harmonised certificate trading as calculated in the EWI study seem largely overestimated"* (page 1). Five reasons are provided for this assessment.

Firstly, the *"EWI study does not define a reference case in line with current realities"* (page 1), as it completely ignores the Renewables Directive and in particular the cooperation mechanisms.

Secondly, EWI 2010 *"overestimates the exploitable potential of best resources across Europe"* as they do not consider grid costs and non-economic limiting barriers. REShaping 2010 points out that the assumption of Ireland increasing the share of renewable electricity from 9% in 2007 to 92% in 2020 or Estonia from 1% to 79% is far beyond imagination. REShaping 2010 estimates shares of about 30% for Ireland and 20% for Estonia to be possible.

Thirdly, they criticize the *"exogenous"* modelling of the technology learning curves, which results in not adequately taking into account that learning needs financing. Consequently, *"the overall investment and generation costs as well as the required support in the harmonised quota scheme are calculated too low"* (pages 1–2).

Fourth, the EWI study's *"calculated savings of generation costs due to a switch to the harmonised quota system (HQS) seem to be largely overestimated"* (page 2) compared to calculations of between 7 and 28 billion € by REShaping 2010.

Fifth and most importantly, the EWI study *"only reflects on investments and generation costs – the decisive policy costs for consumers are ignored"* (page 2). Resch and Ragwitz underline that *"the cumulative 'efficiency losses' resulting from that simplified harmonisation range from 55 to 90 billion €, depending on which study (EWI or futures-e) to rely on"* (page 2).

Building on the shortcomings of EWI 2010 recommendations are developed. *"RE targets can be achieved either by improved (strengthened) national support systems or by a harmonized EU wide support system, as long as support that is offered is technology-specific"* and more explicitly *"RE targets cannot be achieved by a harmonized, technology-neutral support system, because such a system fails to trigger immediate deployment, development and cost reduction of technologies which are currently still more expensive but whose contribution is needed in the mid-to longterm"* (page 5). And it is highlighted that (support) costs for reaching the 2020-targets in a technology neutral scenario – according to their calculations – are considerably higher than in a technology specific approach.

---

[5]More about the project can be found on the project website http://www.reshaping-res-policy.eu/

9.4   2050: LOW-CARBON VERSUS RENEWABLES DEVELOPMENT

The Power-Choice-Scenario was developed and presented in 2009 by the European Union of Electricity Industry[6] (Eurelectric), the lobby organisation of the European electricity producing utilities, with Members in all EU Member States, all being represented in the Board of Directors. For many years, they have been active in Brussels promoting a vision of an energy mix including coal and nuclear and leaving the choice of sources and technologies to the market, which in reality – given the imperfect status of energy markets in the EU and in Member States and elsewhere – resulted in leaving it to the utilities. In recent years this was fine-tuned according to the language of policy communication in Brussels. Eurelectric now aims to *"contribute to the development and competitiveness of the electricity industry, to provide effective representation for the industry in public affairs and to promote the role of a low-carbon electricity mix in the advancement of society"*. This is underpinned by sub-targets of *"Delivering carbon-neutral electricity in Europe by 2050"*, *"Ensuring a cost-efficient, reliable supply through an integrated market"* and *"Developing energy efficiency and the electrification of the demand-side to mitigate climate change"*. The language is in line with sustainability targets now, but the energy sources are still the same, but they are all labelled as *"low-carbon"* energy sources now.

The study is built on this logic – continued use of fossil and nuclear. What is new is significantly including renewables. Starting with grand commitment to ambitious greenhouse gas reduction targets including commitment *"to become carbon neutral by 2050"* and the *"vision of a cost-effective and secure pathway to a carbon-neutral power supply by 2050"* they promise *"deep cuts in carbon emissions from 2025 onwards"* (Foreword by Eurelectric's President of that period, Lars Josefsson). This is not in line with scientific evidence as provided by the IPCC. Relevant greenhouse gas reduction must start much earlier to maintain a chance to meet the target of limiting global warming to 2°C. It is, however, in line with frequent assessment of scientists that Carbon Capture and Storage (CCS) – if ever – will not be economically viable before 2025 or probably later. The scenario foresees CCS as of 2025 (and in a sensitivity analysis for 2035), and new nuclear for about the same period in time – assuming that public debates about nuclear safety will have stopped by then.

To understand and properly assess the study, it is useful to know that the programmatic messages are openly stated at the beginning of the executive summary (page 5). The *"scenario delivers carbon-neutral power in Europe by 2050:*

- *via low-carbon technologies such as renewable energies (RES), carbon capture & storage (CCS) technologies and nuclear power*
- *through intelligent and efficient electricity generation, transmission and use*
- *with intelligent electricity use as the driver for a secure, low-carbon energy future*
- *by promoting the roll-out of electric road vehicles through a drive for widespread energy efficiency in our economy and society*
- *at a lower long-term total energy cost than under the Baseline scenario"*

For modelling the study uses the PRIMES energy model, which is also used by the European Commission, upgraded with recent macroeconomic and power-sector data and assumptions, supplied by Eurelectric's partner organisation VGB Power Tech (page 6). In contrast to a reference scenario the study *"aims for an optimal portfolio of power generation based on an integrated energy market. The PRIMES model calculates the market-optimum, taking into account the technology assumptions developed by the industry"* (page 6). The assumptions listed in detail read like a well-considered mix of commonplace language of the present policy debate and Eurelectric's policy targets for the coming years. *"The scenario assumes that:*

- *Climate action becomes a priority and the EU sets and reaches a target of cutting via domestic action 75% of its CO_2 emissions from the whole economy versus 1990 levels*

---

[6]More about their mission, membership, projects on http://www.eurelectric.org/

- *Electricity becomes a major transport fuel as plug-in electric and hybrid cars develop*
- *All power generation options remain available, including nuclear power in those countries that currently produce it, but envisaged national phaseout policies remain*
- *No binding RES-targets are set after 2020; RES support mechanisms remain fully in place until 2020 and are then gradually phased out during 2020–2030. Energy efficiency is pushed by specific policies and standards on the demand-side during the entire projection period, which will result in slower demand growth*
- *The price of CO_2 ('carbon-value') applies uniformly to all economic sectors, not just those within the ETS, so that all major emitting sectors pay for their emissions*
- *After 2020, an international carbon market defines the price per tonne of CO_2; after 2030 the CO_2 price is the only driver for deployment of low-carbon technologies*
- *CCS technology is commercially available from 2025"* (page 6).

Not having a binding RES-target after 2020 and phasing out support systems is the core objective of Eurelectric's policy activities for the post-2020 era – rather than being an assumption made for this study. Based on these elements, the study provides results like falling primary energy consumption (–20%) resulting in reduced final energy consumption (–30%) compared to the baseline scenario. The study delivers reduced import dependency (–40%) and a *"dramatic"* (page 8) RES-increase to 40% in 2050 (with wind power providing 56% and solar 13% of RES). And after 2025, new nuclear comes into operation, as well as new fossil plants, with CCS. The study shows an increase of total power capacity, attributed mainly to renewable energy. Carbon intensity of Europe's power sector is forecast to be reduced by 66% (page 8). Despite investment of about two trillion Euro$_{2005}$ by 2050, the share of energy costs in percentage of GDP is foreseen to decrease from 10.5% in 2010 via 13% in 2025 to below 10% in 2050. The key outcomes are summarised (page 11), highlighting that the *"major CO_2 reduction in the power sector occurs during 2025 to 2040"* and *"all generation options [. . .] are needed simultaneously"*. To achieve the objectives – and this is back to politics – *"strong and immediate political action is required"*, the first one in the list being *"Technology Choices → Enable the use of all low carbon-technologies . . ."*

From the summary it is already clear that the main objective of the scenario is to support Eurelectric's lobbying activities for continued use of *"all generation options"*, the full study is even clearer on this by including assumptions, which could serve as a back door in case Eurelectric's members fail to deliver even on the post-2025 promises. One of them is the *"Harmonization of carbon policies between the EU and the rest of OECD [. . .] to be completed by 2015. A common target and therefore common effective carbon prices apply within the OECD thereafter"*. Given the reality of climate negotiations and the deplorable status of EU-ETS, this is an extremely optimistic assumption.

For renewables, the approach of strong language (*"dramatically increasing"*) with weak ambition and less binding targets is continued. *"Renewables including biomass double their share in power generation by 2030, relative to 2005, and maintain a share of 40% in 2050"* (page 23). This is claimed to be ambitious, just like the next sentence: *"In the longer term, intermittent renewables contribute roughly 15% to power generation"*. Intermittent is meant to comprise wind and solar power – the word alluding to a high degree of unpredictability and thus risk for security of supply. Most other studies have stopped using this misleading term and replacing it by *"variable"*, which much better describes the quality of wind and solar: it is not always available, but it is predictable and can be balanced through appropriate technologies and market design[7]. In contrast, for fossil and nuclear sources widely discussed problems and barriers are ignored or dismissed. Whereas

---

[7]For more details about integrating variable renewables see IEA 2011 and several other recent studies on this topic. Solutions for the challenge of building an energy system on the logic of (variable) renewables are also discussed a Feature on *"System Transformation"* in GSR2013.

scientific evidence is growing that *"peak oil"*[8] has already been reached or will be reached in the near future, which would result in resource scarcity and increasing prices, the study holds that occurrence of *"peak oil"* in 2020 has a probability of less than 20%, and about 50% in 2030, an assumption based on *"unconventional"* and so far *"undiscovered"* oil reserves, which provide more than 50% of global oil production in 2050 (page 30). Similar deliberations are made for coal and gas.

Disadvantages of fossil and nuclear sources when it comes to costs, particularly on a level playing field with all external costs included in the prices, are downplayed by assumptions about levelized generation costs (pages 40–42). Nuclear power is described to be safe and – despite some cost increases to be expected – all externalities taken into account (page 42): *"The costs of nuclear decommissioning, as provided for in current legislation, were taken into account in the total capital costs. It was verified that nuclear waste treatment associated with further nuclear development to 2050 is manageable; the energy consumed in nuclear fuel treatment and waste management was accounted for in energy balances"*. Nevertheless, levelized costs miraculously stay at around 45 $€_{2005}$ per MWh/net until 2050, whereas the cheapest of the renewable technologies (wind onshore) is assumed to be at 66 $€_{2005}$ per MWh/net. All other renewable energies are set to cost about or only slightly below 100 $€_{2005}$ per MWh/net in 2050 or even far beyond 200 $€_{2005}$ (solar thermal and PV). In contrast, coal and gas – irrespective of whether the carbon price is set to zero or at 30 or 100 $€_{2008}$ – remain significantly cheaper, with coal-CCS being the cheapest choice with 69 € per MWh and a carbon price of 100 €/t in 2050, cheaper than any of the renewables technologies except wind onshore, which is assumed to be at the same cost level.

As a result, it is not surprising that Eurelectric President Josefsson in his presentation at the launch event of the study at the European Parliament (Eurelectric 2009a) had included a slide projecting the electricity mix of 2050 (page 4): RES 38% – nuclear 27% – CCS 30% – other fossil 5%. This clearly underlines Eurelectrics policy objective of not having a new binding renewable energy target for 2030 and trying to replace it by a decarbonisation target or a greenhouse gas reduction target. Such an approach would be in line with continued use of nuclear and continued belief in CCS taking up some time soon. For renewables development that would be devastating.

9.5   2050: REACHING 80% GREENHOUSE GAS REDUCTION AND MORE

With the *"Roadmap 2050 Practical Guide to a Prosperous Low Carbon Europe"* (ECF 2010) the European Climate Foundation[9] presented a three volume study aiming to break *"new ground by outlining plausible ways to achieve an 80% reduction target from a broad European perspective, based on the best available facts elicited from industry players and academia, and developed by a team of recognized experts rigorously applying established industry standards"* (ECF 2010, Volume 1, page 3). Volume 1[10] provides *"Technical and Economic Analysis"*. Volume 2[11] is the *"Policy Report"* and volume 3[12] is the *"Graphic Narrative"*. The intention is to *"initiate a critical policy debate across Europe relating to the future of energy markets"* (Volume 2, page 6). A highlighted result is the technical and economic feasibility of an 80% greenhouse gas reduction by 2050 without relying on international carbon offsets through various scenarios, all of which are not significantly more costly than the business-as-usual case.

---

[8] *"Peak oil is the point in time when the maximum rate of petroleum extraction is reached, after which the rate of production is expected to enter terminal decline"* (Wikipedia, viewed 10.3.2013).

[9] The European Climate Foundation (ECF) was established in 2008 to promote climate and energy policies that greatly reduce Europe's greenhouse gas emissions. More about their work and funding partners can found seen at http://www.europeanclimate.org/

[10] Elaborated by McKinsey & Company, KEMA and The Energy Futures Lab at Imperial College London; Oxford Economics and the ECF.

[11] Elaborated by E3G, The Energy Research Centre of the Netherlands and the ECF.

[12] Produced by The Office for Metropolitan Architecture and the ECF.

Table 9.3.   The Assumed technology mix in the analysed scenarios (in %).

	Coal*	Gas**	Nuclear	Wind onshore + offshore	PV	CSP	Biomass	Geothermal	Large hydro
100% RES	0	0	0	15 + 15	19	5 + 15***	12	2 + 5****	12
80% RES 10% CCS 10% nuclear	5	5	10	15 + 15	12	5	8	2	12
60% RES 20% CCS 20% nuclear	10	10	20	11 + 10	12	5	8	2	12
40% RES 30% CCS 30% nuclear	10	10	20	11 + 10	12	5	8	2	12
Baseline: 34% RES 49% coal/gas 17% nuclear	21	28	17	9 + 2	1	1	8	1	12

Data from Exhibit 12 (ECF 2010, Volume 1, page 50) and page 77.
*Only with CCS or CCS-retrofit in decarbonisation scenarios – without CCS in Baseline.
**Only with CCS in decarbonisation scenarios – without CCS in Baseline.
***Import from North Africa.
****Enhanced geothermal.

The Roadmap – using a back-casting methodology (deriving pathways from a set target, i.e. at least minus 80% GHG-reduction in 2050) – underlines the feasibility of the required decarbonisation while maintaining *"supply reliability, energy security, economic growth and prosperity"* (Volume 1, page 6). An important finding of an initial analysis is *"that it is virtually impossible to achieve an 80% GHG reduction across the economy without a 95 to 100% decarbonized power sector"* (Volume 1, page 6). The Roadmap is one of the first among several studies published in 2010 und 2011, which were examining pathways towards far reaching greenhouse gas reductions and/or very high shares of renewable energy. Some of the other key publications are presented later in this chapter.

Without further explaining why, *"three different decarbonized power sector pathways"* are presented in the Roadmap and compared to a BAU-scenario (for details see Table 9.3), which differ in the respective shares of renewable energy, nuclear power and CCS (40, 60 or 80% renewables combined with 30, 20 or 10% of nuclear and CCS). In addition, a scenario with 100% electricity from renewables is *"assessed, primarily on the dimension of maintaining the acceptable level of service reliability"*. The approach of the study to use *"by design a mix of technologies [. . .] to avoid over-reliance on few 'silver bullet' technologies"* cannot explain why the 100% RE scenario is not fully examined – renewables alone are already a mix of various technologies and resources. The report also falls short of explaining why the 100% RE scenario has to include 15% energy imports from CSP in North Africa.

The study finds that ambitious energy efficiency measures are important to facilitate the decarbonisation process. The earlier the transition starts, the less expensive it will be. A ten year delay in implementing major decisions will increase the necessary annual investment from 65 billion € in 2025 to more than 90 billion in 2035. The necessary scaling-up of the supply chain will be easier the earlier the process starts.

The study sets safety margins by not including technologies which are not mature or at least in a late development state today. New groundbreaking technologies would therefore increase the viability of the scenario and decrease the costs. In any case, new power lines all over Europe are

Table 9.4.   Assumed necessary back-up capacity (in GW installed in 2050).

	Fossil fuels	Solar PV	Wind onshore	Wind offshore	Other	Back-up
80% RES 10% CCS 10% nuclear	80	815	245	190	420	270
60% RES 20% CCS 20% nuclear	155	555	165	130	455	240
40% RES 30% CCS 30% nuclear	240	195	140	25	490	190
Baseline: 34% RES 49% coal/gas 17% nuclear	410	35	140	25	380	120

Data from Exhibit 13 (ECF 2010, Volume 1, page 51).

necessary to fully tap the calculated potentials, as well as *"aggressive energy efficiency measures"* (Volume 1, page 10). The Roadmap sees increasing capital needs for the decarbonisation pathways, but decreasing energy costs per unit of GDP, due to less fossil fuel needed and due to increased efficiency, amounting to lower costs of 350 billion € per year by 2050 or 1,500 € per year and household (Volume 1, page 11). A positive effect of 300,000 to 500,000 new jobs by 2020 is expected by implementing the decarbonisation scenarios compared to the baseline, whereas 250,000 jobs could be at risk in all scenarios, baseline and decarbonisation.

Implementation of the decarbonisation scenarios is seen as a major challenge, because about 5,000 square kilometres are needed for solar panels until 2050 (half of it on roof tops) and 100,000 wind turbines of up to 7–10 MW capacity (half of them offshore) have to be newly built or replacing old ones, which translates into 2,000 to 4,000 turbines per year (Volume 1, page 14). In addition, a much higher degree of interconnection will require new transmission capacity with several thousand kilometres of new power-lines. All interconnectors between EU Member States need to be enhanced, some of them to a very large extent. For example, the study assumes a needed transmission capacity between France and the Iberian Peninsula of 15 to 40 GW – up from 1 GW today, with the high end for the 80% renewables option.

Another critical point arising from the Roadmap is the assumed need for installed back-up generation plants in the different scenarios, all running at very low capacity factors of less than 5% in the 40, 60 and 80% renewables scenario – and 8% in the 100% scenario. Without convincing evidence, the study assumes that 10 to 15% of the whole generation capacity will be needed as back-up capacity, the high end again for the 80% renewables scenario. As a result, up to 270 GW of back-up plants are assumed to be necessary, compared to 120 GW in the reference case (Table 9.4). Based on these assumptions the 100% scenario becomes considerably more expensive than the other scenarios – with levelized costs of energy being 10% higher than for the 80% iterations.

Following more recent discussions, this high amount of back-up capacity running at very load factors is not necessarily needed and probably not the smartest solution, if grid infrastructure, storage capacities, flexibility options, demand response and a smart mix of centralized and decentralized technologies are combined, and even more so, if the electricity sector is interconnected with the heating and cooling sector as well as electromobility with plug-in EVs serving as storage and balancing options.

The study holds that *"to put the EU-27 on a path to 80% GHG reductions by 2050, a 20 to 30% reduction must be realized in 2020"* (Volume 1, page 89) – a wise remark en passant, avoiding the ongoing debate about whether the EU should increase the ambition level to minus

30% in 2020. Apart from proving the technical and economic feasibility of the scenarios, the Roadmap provides some advice on necessary steps to be taken in the near futures. Five priorities for the next five to ten years are highlighted. Energy efficiency must play an important role. The *"low carbon technologies"* need to be developed and deployed. Grid and market integration needs to be implemented. Fuel shift in the transport and in the building sector will accelerate the decarbonisation pathways. Markets need to be developed to deliver the necessary investment.

Eventually, some *"delivery risks"* are assessed, which might impede implementation of the decarbonisation scenarios. For the 40% renewables (plus 30% nuclear and 30% CCS) scenario, technology risks and social acceptance of nuclear and CCS as well as fuel resource depletion *"may constrain the pathway in the long term (beyond 2050) and may cause spikes in the energy price in the shorter term"* (Volume 1, page 89). These risks can certainly be avoided by choosing the 80% (or higher) renewables pathway, where mitigating the main delivery risks (technology development and learning curves, building supply chains, *"resolving the intermittency issue"* and *"overcoming local opposition"* to large scale installations) is probably much easier than for CCS and nuclear. A joint risk for all pathways is seen in insufficient efficiency development and lacking policy support.

Volume 2 provides policy recommendations to answer the challenges and facilitate the transition. *"It highlights where action is required at the EU level and where it is most appropriate to be progressed within Member States or regions. The objective of these proposals is to 'kick-start' the conversation amongst policy makers, regulators, investors and other stakeholders that is required if we are to deliver changes that are robust and timely"* (Volume 2, page 7). They are referring to concrete legislative acts that should be undertaken, such as revising the Energy Services Directive, the Eco-Design Directive, complementing and reviewing the ETS-Directive and much more (Volume 2, pages 9–11).

In short, the report is asking for a binding energy efficiency target on EU-level, an update of the emissions trading scheme, a *"new 'Climate and Resources'framework"* beyond 2020 and up to 2050, budget allocation to ensure funding and expansion of ACER's and ENTSO-E's mandate. Member States are requested to develop and *"adopt aggressive targets and strategies for the deployment of energy efficiency measures that will double to triple the current rate of overall energy efficiency improvement"* (Volume 2, page 8). Furthermore, they shall consider new renewables targets beyond 2020, implement market adaptations, regional cooperation with neighbouring countries, review the mandate of the regulators, establish timetables and ensure funding.

9.6   THE INDUSTRY'S PERSPECTIVE: 100% RENEWABLE ENERGY IN 2050

In contrast to ECF 2010, the European renewable energy sector (and some environmental NGOs) decided to focus rather on the potentials of renewable energies – and the benefits they could provide for growth, innovation, jobs and environment, including climate protection. This approach was a major driver for the European Renewable Energy Council (EREC)[13] and member associations to present *"REthinking 2050 – A 100% Renewable Energy Vision for the European Union"* (EREC 2010a) in April 2010.

In the foreword to the study, EREC's longtime President, Arthouros Zervos, listed the challenges to be met. And he concluded: *"RE-thinking 2050 outlines a pathway towards a 100% renewable energy supply system by 2050 and clearly shows that it is not a matter of technology, but rather a matter of making the right choices today to shape tomorrow"*. The study describes potentials and possible growth pathways of renewable energies via 2020 and 2030 to 2050. It comprises two different scenarios for consumption development, one of them assuming *"what might be possible in 2050 on the basis of an aggressive efficiency and energy savings approach"* (page 21). This would result in energy savings of 38% compared to today's levels.

---

[13] More about the European umbrella organization of the renewables sector at http://www.erec.org/

Table 9.5.   Renewable electricity: installed capacity [GW].

	2007	2020	2030	2050
Wind	56	180	288.5	462
Hydro*	102	120	148	194
PV	4.9	150	397	962
Biomass	20.5	50	58	100
Geothermal	1.4	4	21.7	77
CSP	0.011	15	43.4	96
Ocean	–	2.5	8.6	65
Total RES-E capacity (GW)	**185**	**521.5**	**965.2**	**1,956**

Source: EREC 2010a, page 23.
*excluding pumped storage.

Table 9.6.   Renewable heating and cooling consumption [Mtoe].

	2007	2020	2030	2050
Biomass	61.2	120	175	214.5
Solar thermal	0.88	12	48	122
Geothermal	0.9	7	24	136.1

Source: EREC 2010a, page 27.

For the electricity sector, the study provides an average annual growth rate of installed renewables capacity of 14% up to 2020 and 8.5% until 2030, a trend that is supposed to continue until 2050. Regarding the individual technologies, wind is predicted to be the strongest growing until 2020, being overtaken by PV by 2030. Depending on consumption assumptions and on the installed capacities shown in Table 9.5, the shares of renewables in electricity consumption will increase to 39.2–39.8% in 2020, 65–67% in 2030 and 100–143% in 2050 (page 25).

The calculations include a steep increase of electricity demand after 2030 due to increased use of heat-pumps and due to the impact of electric vehicles, which are expected to have major growth rates then.

The heating and cooling sector, which accounts for 49% of the EU's final energy demand today, will remain to be dominated by biomass use, but solar thermal and geothermal heating are expected to grow considerably after 2020. In 2050, the biomass share could fall below 50%. Table 9.6 shows the renewables contribution to heating and cooling consumption. Based on this output and depending on the different consumption scenarios, the share of renewables in the sector would be 28–29% in 2020, 52–57% in 2030 and 100–143% in 2050 (page 28).

For the transport sector, today's 98% dependency from oil will have to be overcome by increased efficiency, expansion of biofuels and biomethane as well as new technologies, such as hydrogen, hybrid vehicles and in particular electric vehicles. The study foresees biofuels consumption to increase from 7.88 Mtoe in 2007 via 34 Mtoe in 2020, 44.5 Mtoe in 2030 to 102 Mtoe in 2050, which would result in biofuels shares in the transport sector of 8.7–9% in 2020, 11.4–12% in 2030 and 68.6–98% in 2050 (page 30). In 2050, renewable electricity could contribute more than half of transport fuel demand.

Based on the assumptions and calculations for the three sectors and the different efficiency scenarios, the share of renewables in Europe's total final energy consumption could increase to 100% in 2050 (for details and variations see Table 9.7).

After having assessed potentials of the different sources the study calculates related benefits for supply security, economy and environment. Based on oil price assumptions of 82 US$ per barrel in 2010, 100 US$ in 2020, 120 US$ in 2030, 150 US$ in 2040 and 200 US$ in 2050 potential cost savings are calculated. Hence, in 2020, fossil fuel costs of € 158 billion could be avoided, € 325 billion in 2030 and € 1,090 billion in 2050 (pages 36–37).

Table 9.7. Contribution of renewable energy to final energy onsumption.

	2020	2030	2050
Total share of RES [%]	24.5–25.5	42.4–44.4	96–137

Source: EREC 2010a, page 32.

Another element of imbalance between renewables and conventional energy is the internalisation of external costs. Traditionally, energy customers are paying indirectly – via taxes, insurance, social contributions – for the impact of energy on life, health and security. With external costs included, the study calculates that by 2020 an average mix of renewables will start to be cheaper than the fossil and nuclear mix. By 2040 the very latest, this would even be the case without internalising all external costs (page 38).

The study provides calculations about the development of capital costs for the different technologies and resulting investment up to 2050. Accordingly, by 2020 the total cumulative investment in renewables will amount to € 963 billion, about 70% of it going into the electricity sector and 27% to heating and cooling. Until 2030, € 1,620 billion will have been invested, 55% for electricity and 42% for heating and cooling, going up to € 2,800 billion in 2050, 51% for electricity and 46% for heating and cooling (page 40).

Among the social benefits, employment effects are highlighted, indicating that about 6.1 million jobs could be created in Europe's renewables sector by 2050, up from 550,000 in 2009 via 2.7 million in 2020 and 4.4 million in 2030 (page 43).

Deployment of renewables according to EREC's calculations would result in greenhouse gas reduction of around 95% below 1990 levels and thus be fully in line with the target of limiting global temperature increase to a maxim of 2°C until the end of the century. Based on a CO_2-price of 41 €/t, the total benefit in 2020 could be € 492 billion, increasing to € 3,800 billion by 2050 with a CO_2-price of 100 €/t (page 42) – calculations which are highlighting the need for creating significant and stable carbon prices through appropriate instruments. The present debate about saving the EU-ETS to deliver on this objective is of high importance, but a positive outcome cannot be taken for granted.

As stated in the foreword, the scenario should not be taken as a prediction about an inevitable future development, but it is meant to offer a sustainable pathway – provided that enabling policy decisions are taken in due time. In the final chapter, the study summarizes the most important policy recommendation for achieve the objectives and reaping the benefits.

First and foremost, a strong commitment of the European Union and its Member States is needed. The 2020-framework, in particular the Renewables Directive, is a good starting point, but more needs to be done.

The following ambitious and unambiguous policy decisions are considered to be necessary to achieve the goal of a renewables based, sustainable energy supply by 2050:

- an ambitious framework for Europe's energy demand,
- effective and full implementation of the Renewables Directive in all Member States,
- binding renewable energy targets for 2030,
- full liberalisation of the energy market,
- phasing out all subsidies for fossil and nuclear energy and introducing an EU-wide carbon tax.

In addition, infrastructure development and enabling measures for consumers to play their role need to be agreed and implemented.

9.7 2050: OUTLINING THE ENERGY [R]EVOLUTION

Another example of a scenario, which aims at elaborating how 100% renewable energy can be achieved in Europe, is Energy [R]evolution 2010, which was derived from a global scenario for all parts of the world which was launched by Greenpeace International in 2007 (Energy [R]evolution 2007). Since then, the scenario has been further refined and developed. The European scenario was presented in cooperation with EREC in 2010, underpinning by more in-depth calculations and assumptions the policy goals of Europe's renewable energy sector and most of the environmental NGOs.

The scenario is embedded in the global policy debate, where Energy [R]Evolution has become a respected contribution to policy development, although – or maybe just because – it is the most ambitious of all global scenarios (in terms of greenhouse gas reduction and renewables deployment) analysed in IPCC 2010. It outlines two pathways, a basic scenario and an advanced scenario, both mainly using existing and proven technologies and a broad mix of various renewables. General parameters, such as population development and GDP are based on the reference scenario of WEO 2009, which was extrapolated from 2030 to 2050 by the German Aerospace Center (DLR).

The study builds on five *"key principles"* (page 11):

1. *"respect natural limits – phase out fossil fuels by the end of this century on the global level and by 2050 on the European level [. . .],*
2. *equity and fairness [. . .],*
3. *implement clean, renewable solutions, decentralize energy systems [. . .],*
4. *decouple growth from fossil fuel use [. . .],*
5. *phase out dirty, unsustainable energy [. . .]".*

The basic scenario reduces the EU's CO_2-emissions by 80% until 2050, while phasing out nuclear by 2030. The advanced scenario reaches reductions of 95% and *"dramatically improves energy security, boosts green technology leadership"* (page 7). The scenario cuts the lifetime of coal-fired power plants to 20 years and implies a faster uptake of efficient combustion vehicles and electromobility, as well as more CHP and electrification for process heating.

The authors advocate *"five policy tasks to start the energy [r]evolution"* (page 9).

1. *"Develop a vision for a truly sustainable energy economy for 2050 that guides European climate and energy policy [. . .]*
2. *Adopt and implement ambitious targets for emissions reductions, energy savings and renewable energy [. . .]*
3. *Remove barriers [. . .]*
4. *Implement effective policies to promote a clean economy [. . .]*
5. *Redirect public finance [. . .]".*

The advanced scenario would require € 3.8 trillion of investment in the EU until 2050 compared to around € 2 trillion for the basic scenario. On the other side of the equation, fuel cost savings amount to € 2.1 trillion in the basic scenario and € 2.6 trillion in the advanced scenario. Starting in 2040, fuel cost savings would fully cover all investment costs including CHP and back-up capacity (page 8). Until 2030 1.2 million jobs would have been created in the advanced scenario, compared to 420,000 in the reference scenario.

The authors underline that decisions have to be taken in due time. *"Within the next ten years, the power sector will have to decide"* (page 12), whether to invest in renewables or in fossil and nuclear. To make the renewables path happen a transition phase will be needed for implementing the necessary, more decentralized infrastructure including gas-CHP as a bridging technology. Virtual power plants will have to be established, storage options need to be set up.

Like EREC 2010a, the authors highlight the need for political commitment of the European Union and its Member States to full and ambitious implementation of the existing 2020 targets, developing suitable policies for further increasing ambitions and the willingness to lead by example.

As results of the calculations the share of renewable energy in the European Union would increase considerably until 2050. In the advanced scenario, renewables contribute 91.7% to the

Table 9.8.   Shares of renewables in 2050*.

	Reference [%]	Basic [%]	Advanced [%]
Final energy demand	23.8	61.8	91.7
Electricity generation	40.6	88.5	97.3
Heat supply	22.7	55.9	92.2
RES share transport	9.5	48.1	85.9

*Source: [R]Evolution 2010, pages 62–64.

total final energy consumption, compared to 61.8% in the basic scenario and only 23.8% in the reference scenario.

As an overall result it is shown how a nearly complete supply with renewable energy could be achieved, with the hurdles being significantly higher in the transport sector than in heating and cooling and in power production. Consequently, in all three scenarios the highest share of renewables is expected in the electricity sector, the lowest share in the transport sector, a tendency that can be observed in all analysed scenarios. Figures about the calculated shares of renewables in 2050 can be found in Table 9.8.

9.8   EU COMMISSION ROADMAPS 2050: VERY HIGH SHARES OF RENEWABLE ENERGY

In 2011, the European Commission presented three energy related strategic papers, each labelled as a *Roadmap*. The first in the row, published on March 8, was the Low Carbon Roadmap, the *"Roadmap for moving to a competitive low carbon economy in 2050"* (European Commission 2011a), which outlined the challenges and opportunities of reducing greenhouse gas emissions by 80–95% in 2050, in particular highlighting benefits of increasing the 2020 greenhouse gas reduction target from 20% to 25 or 30%. On March 28, the Transport Roadmap, the *"Roadmap to a Single European Transport Area – Towards a competitive and resource efficient transport system"* (European Commission 2011d) followed, outlining ways to reduce greenhouse gas emissions in the transport sector by 60% until 2050. Eventually, on December 15, the *"Energy Roadmap 2050"* (European Commission 2011f) concluded the series of roadmaps, assessing different *"decarbonisation scenarios"* to deliver on the reduction targets defined by the Low Carbon Roadmap.

All three Roadmaps were meant to and actually did open discussions on how to proceed towards meeting the ambitious decarbonisation levels, which are necessary to keep the window open for limiting global warming below 2°C. Apart from providing cost and technology pathways, they also opened discussions about new milestones for bridging the gap from the agreed 2020 targets to the tentative and non-binding 2050 visions, thus taking on board – at least as a series of questions to be analysed, assessed and decided – the strategic demands from NGOs regarding climate change and from the renewable energy sector regarding stable and reliable framework conditions on a mid-term basis.

In May 2011, the European Renewable Energy Council had launched a campaign for a new binding renewable energy target of 45% in 2030, which was underpinned by a new publication in April 2013 (EREC 2013b). The campaign is foreseen to continue at least until a 2030-framework is agreed in 2014 or in 2015. This new publication and other discussions about the post-2020 framework are analysed in Chapter 11 of this book.

9.8.1   *Low Carbon Roadmap 2050*

Building on the proposed *"flagship initiative of the Europe 2020 Strategy"* for a *"resource-efficient Europe"*[14] and the *"Europe 2020 Strategy for smart, sustainable and inclusive growth"*

---

[14]http://ec.europa.eu/resource-efficient-europe/

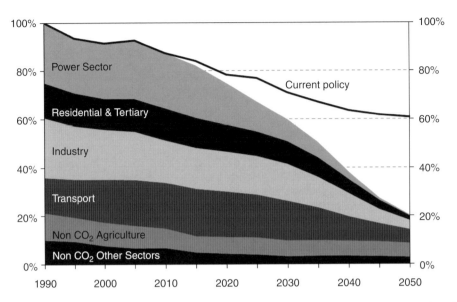

Figure 9.1.   EU Greenhouse gas emissions towards an 80% domestic reduction (100% corresponds to the emissions of the year 1990).

(European Commission 2010a), which relate to the 2020 targets, the Roadmap finds that the *"EU is currently on track to meet two of those targets, but will not meet its energy efficiency target unless further efforts are made"* (page 3). Consequently, milestones are outlined, *"which would show whether the EU is on course for reaching its target, policy challenges, investment needs and opportunities in different sectors, bearing in mind that the 80 to 95% reduction objective in the EU will largely need to be met internally"* (page 3).

The roadmap calculates milestones towards 2050, finding that the most cost effective way for the Union to achieve domestic greenhouse gas reductions of 80% until 2050 without offsetting part of it from the international carbon market would be aiming at reductions of 40% in 2030, 60% in 2040 and 80% in 2050. This would result in a reduction of 25% in 2020, as shown in Figure 9.1 (from European Commission 2011a), which also highlights that current policies would fail to meet the 80% reduction target. A 25% reduction by 2020 could be reached, if the EU delivered on all three 2020-targets, including the efficiency target, which will not be possible, however, without additional measures. The analysis also shows that less ambitious pathways would result in *"significantly higher overall costs over the entire period"* (page 5), due to lock in of carbon intensive investments and resulting higher carbon prices.

The roadmap reiterates the key role of almost totally decarbonising the electricity sector by 2050, translating this into increasing the *"share of low carbon technologies in the electricity mix"* to *"nearly 100% in 2050"*, up from *"around 45% today to around 60% in 2020, including through meeting the renewable energy target, to 75 to 80% in 2030"* (page 6), which seems reasonable given the need for full decarbonisation, but which raises concerns about opening backdoors for undesirable developments. Not even for making the case for decarbonisation can it make sense to simply add up renewables and nuclear (and later CCS) to *"low carbon technologies"*. From a strategic point of view, nuclear and coal power plants are not only politically inacceptable and environmentally harmful, they also delay the transition towards decarbonisation due to their inflexibility, leading to either grid congestions and/or higher costs for new power lines, which would not be needed (or needed to be designed differently) without inflexible power plants blocking grid capacity. However, the low carbon narrative is a constant element of EU communications on greenhouse gas reduction. Initiatives for creating some kind of support for nuclear using the low carbon pretext were and are launched repeatedly.

On the positive side, the Low Carbon Roadmap finds that the EU-ETS will be critical for driving decarbonisation of the electricity sector. The need for a long-term predictable and sufficiently high carbon price is highlighted, which may require reforms of the instrument and additional tools, such as energy taxation. The crucial role of demand side efficiency and particularly the *"significant use of renewables"* (page 7) is emphasized.

For the transport sector, the roadmap foresees a critical role of improved fuel efficiency, at least until 2025. In parallel, sustainable biofuels and other technologies like plug-in hybrids and electric vehicles need to be encouraged. The roadmap advocates the development of 2nd and 3rd generation biofuels so that they can play in important role – only after 2030.

For the built environment the energy performance of buildings needs to be drastically improved and *"emissions in this area could be reduced by about 90% by 2050"* (page 8). To achieve this, like in the transport sector, innovative financing models will be needed and including electricity as a source will be important.

For the industrial sector, significant greenhouse gas reductions of 83 to 87% are foreseen, to be achieved by advanced equipment and efficient processes. Carbon capture and storage – on a broad scale after 2035 – is seen as an indispensable means to *"capture industrial process emissions"* (page 9).

For the agricultural sector, non-CO_2 emissions should be reduced by 42 to 49% until 2050, mainly through improved practise. And of course, the issues of direct and indirect land use change need to be addressed and solved.

The roadmap calculates annual fuel cost savings between € 175 billion and € 320 billion – figures which look nice, but do not indicate how high the savings might really be as long as they include expensive nuclear and CCS as well as increasingly cheap renewables technologies. New jobs are predicted in the range of 1.5 million by 2020, which – compared to the 2.7 million of EREC 2010 – is not very impressive and may be seen as an indication how small the number of jobs in nuclear and CCS would actually be.

9.8.2   *Transport Roadmap 2050*

The White Paper titled *"Roadmap to a Single European Transport Area – Towards a competitive and resource efficient transport system"* (European Commission 2011d) contains a vision for the full range of questions around the transport sector, including passengers and goods, road, sea and air, transport security and passenger rights, networks and level playing field, urban transport, and also greenhouse gas reduction and energy use.

For achieving the 60% greenhouse gas reduction target, the White Paper lists 10 goals of highlighted importance (pages 9–10) – among them the shift to non-conventional fuels for cars, carbon free city logistics, cleaner fuels for aviation, modal shift away from road transport, transport management, and better navigation, as well as the strict application of "user pays" or "polluter pays" principles.

9.8.3   *Energy Roadmap 2050*

The most controversially discussed and most quoted of the roadmaps seems to be the *"Energy Roadmap 2050"* (European Commission 2011f), which was developed to show pathways how the emissions reduction goals of the Low Carbon Roadmap could be achieved, *"while at the same time ensuring security of energy supply and competitiveness"* (page 2). The study finds that existing policies designed for 2020 will continue to deliver results afterwards, reaching around 40% greenhouse gas reduction by 2050, which is by far not sufficient to comply with greenhouse gas reduction needs.

The authors see opportunities for tackling the challenges, because *"in this decade, a new investment cycle is taking place, as infrastructure built 30–40 years ago needs to be replaced"*. On the other hand, *"there is inadequate direction as to what should follow the 2020 agenda"*. Different routes are therefore explored, all except the reference scenarios meeting the decarbonisation

Table 9.9.   Energy Roadmap 2050 – Scenario overview*.

Current trend scenarios	
Reference scenario	Current trends and longterm projections. Policies adopted by March 2010, including 2020-targets and EU-ETS
Current Policy Initiatives (CPI)	Updates adopted measures, including Energy 2020 strategy and Fukushima nuclear accident, proposed actions of "Energy Efficiency Plan" and of "Energy Taxation Directive"
Decarbonisation scenarios	
High Energy Efficiency	Energy demand 41% lower in 2050. Stringent minimum requirements for buildings and appliances. High renovation rates of buildings
Diversified supply technologies	No preferred technology. No support systems. Decarbonisation driven by carbon pricing. Public acceptance of nuclear and of CCS
High Renewable Energy Sources (High-RES)	Strong support for RES. 75% in 2050 in gross final energy consumption. 97% of electricity consumption
Delayed CCS	Similar to Diversified supply scenario, but assuming CCS to be delayed, leading to higher shares of nuclear energy
Low nuclear	Similar to Diversified supply scenario, but assuming no new nuclear due to lack of public acceptance. Instead, higher penetration of CCS (32% in power production of 2050)

*Sources: European Commission 2011f, European Commission 2011f.

targets of at least 80% by 2050, and for modelling reasons all reaching similar reduction rates. Like in the Low Carbon Roadmap, all *"low carbon technologies"* are potential solutions, including unconventional sources like shale gas and including controversially debated nuclear and CCS. The aim is described as to *"develop a long-term European technology-neutral framework"* (page 3), including assessment about potential cost savings through a *"more European approach"*.

The Energy Roadmap 2050 analyses seven scenarios, two of them reference cases and five of them meant to deliver the required decarbonisation. An overview with short description of the scenarios can be found in Table 9.9, more details in the Impact Assessment accompanying the roadmap (European Commission 2011h).

The choice of scenarios is not fully justified or explained, and it is only partly evident. For the two current trend iterations, the basic settings seem to be acceptable, just like for the high efficiency case and the diversified supply case, although the latter is closer to wishful thinking than to reality, when it comes to deployment of CCS and new nuclear. The idea of having a High-RES scenario is convincing as such, but no reasons are given for complete omission of a 100% renewables scenario, which would have been a logical reaction to the various scenarios published in 2010 and described earlier in this chapter. And it is neither evident nor consistent with policy development in the EU in recent years that the Commission does not assess a scenario combining high ambitions of the two major pillars energy efficiency and renewables, for example by assessing a *High-RES & high-efficiency* case.

The omissions are even more astounding compared to the arbitrary settings for the *"low nuclear"* and the *"delayed CCS"* scenarios. Neither does the roadmap provide any scenario without nuclear (at least assuming a gradual phase-out until 2050), nor a scenario without commercial uptake of CCS, and even less so a scenario without both of these options. Instead, the two scenarios are strangely interlinked by assuming that either nuclear or CCS will be widely available and publicly accepted, not even considering to displace nuclear and/or CCS by more renewables and/or higher efficiency. And it should at least be mentioned that all scenarios are very much focussing on the power sector and only marginally touching heating and cooling and transport.

Given these weaknesses and other inconsistencies, e.g. regarding cost decreases which seem underestimated for renewables and overestimated for CCS, or grid costs solely attributed to renewables, the Roadmap nevertheless provides some useful results for upcoming policy decisions.

Major findings from the Commission's point of view, including some that need to be questioned for the sake of clean and sustainable energy development in the EU, are summarised as *"Ten structural changes for energy system transformation"* (pages 5–8). The following developments are observed or predicted:

- Decarbonisation is possible and in the long run can even be less costly then the reference cases.
- Higher capital expenditure will be met by lower fuel costs, including reduced import dependency.
- Electricity plays a much greater role in all scenarios.
- Electricity prices will increase until 2030 and then decline (except for the high-renewables case, where they continue to increase (for assessment of this flawed exception see below).
- Household (and SME) expenditure for electricity will increase, which can (partly) be offset by efficiency gains.
- Energy savings are crucial.
- Renewables rise substantially in all scenarios.
- CCS has a pivotal role to play.
- Nuclear energy provides an important contribution.
- Decentralisation and centralised systems increasingly interact.

Indeed, all five decarbonisation scenarios foresee very high shares of renewables in 2050, the lowest (54.6%) in *"diversified supply"* and the highest (75%) in *"High-RES"*. For 2030, however, even the *"High-RES"* scenario only provides a 31.2% share in final energy consumption, which is significantly below industry estimations, and which would mean drastically reducing growth rates after 2020. In the electricity sector, even the lowest pathways as in diversified supply foresee nearly 60% of renewables in 2050 – and already more than 51 in 2030, which is another strange result, as this would mean a dramatic slowdown of renewables growth after 2030.

All scenarios, including the reference cases show similar cost developments, the average annual total energy system costs being extremely close to each other, with the *"low CCS"* showing the lowest costs after 2030 and the efficiency case the highest, even more than the *"High-RES"*. The latter is burdened by biased assumptions, e.g. too high costs for PV compared to actual learning curves, strangely low capacity factors for nuclear (50%), coal and lignite (20%), and natural gas (24%) in 2050. Such capacity factors would actually imply that nuclear power plants either reach a much higher flexibility than they do today and/or that investors for new power plants could be found despite these capacity factors. The same applies for coal and lignite. In 2050, none of the present installations will still be operational. And in the meantime, investment in coal would certainly not take place to a relevant extent under the scenario assumptions. As a result of the low capacity factors together with grid extension costs (they increase dramatically after 2030 in the *"High-RES"* case) and reserve capacity calculations and storage need (also dramatically increasing in the renewables case after 2030), the *"High-RES"* case is the one with the highest – and constantly increasing – electricity prices after 2030. With more realistic assumptions, the result could easily have been the opposite.

There is another remarkable result in the Roadmap which is worth mentioning. Import dependency is the lowest in the *"High-RES"* scenario, which could be explained by the fact that renewables are resources which need not be imported, but are available domestically, and in the case of wind and solar, even for free.

9.9   AFTER THE ROADMAPS: STRIVING FOR 100% RENEWABLES IN 2050

While the discussions about the three roadmaps are going on in the European Parliament and in the Council and while they will certainly not be unanimously concluded when this book is printed, a more precise view on how to combine existing targets and the state of implementation with drafting next steps for moving towards 2050 is needed. Alternatives need to be considered regarding the interaction between efficiency measures and renewables deployment. Criteria need to be further refined how to choose between different renewable energy technologies and sources.

The European Section of WWF published a proposal to this end in February 2013, titled *"Putting the EU on Track for 100% Renewable Energy"* (WWF 2013) and specifying that this should be reached by 2050 the latest and therefore served as a starting point for the study's assumption setting and calculation methodology.

Among the key findings (page 3) the *"lack of political ambition"* and the *"need for greater clarity on policy frameworks for renewable energy and energy efficiency"* are highlighted. With suitable frameworks in place, in 2030, the EU could use 38% less energy compared to business as usual projections, source more than 40% of the energy demand from renewables and reduce energy related greenhouse gas emissions by 50% compared to 1990 levels. The Union would thus be on track for 100% renewables in 2050.

The study *"shows what has to be achieved in Europe over the next two decades in order to keep the option of a fully renewable global energy system within sight"* (page 7). It undertakes the innovative approach to combine ambitious energy savings with scaling up renewables in the order of sustainability of the different sources and technologies, thus starting with solar, wind, water followed by geothermal and using bio-energy only under very strict conditions.

Following the recommendations of WWF, in particular increasing the greenhouse gas reduction target for 2020 to 30%, is expected to generate up to 6 million jobs in the EU by then, a million more than with the 20% target. This would also reduce fossil fuel costs by more than € 500 billion annually.

The study provides detailed recommendations for the different sectors for reaching the 2030 objectives. For the industry, this would include achieving energy intensity for aluminium, cement, steel and paper production of 60 to 70% of today's levels by more recycling, ambitious refurbishment and strict BAT application. For other sectors, energy supply, processes and motor driven systems need to become more efficient. For the building stock, retrofitting rates of 2.5% annually are foreseen; ambitious insulation is required, just like 25% of heat demand needs to be sourced from solar thermal or from heat pumps. As a result, 45% of the stock should be retrofitted by 2030. For new buildings, which will have to be powered by electricity only, near zero energy standards will be required. Eventually, for the transport sector, the energy intensity shall be reduced to 60% for passenger transport and 70% for freight transport compared to 2000 levels. This could be achieved by more efficient engines, high rates of electrification and by use of sustainable biomass. Altogether energy demand is drastically reduced until 2030.

The share of renewables in 2030's electricity supply is calculated to be 65%. For assumed grid capacity restrictions this is combined with a cap of 45% in annual average from variable sources (wind and solar – page 15). Another limiting factor for resources is the availability of sustainable biomass, where legally binding criteria are deemed to be necessary in the EU (and beyond). These criteria (page 20) include full accounting of carbon emissions, zero use of valuable land, sustainable forest management, respecting human rights and credible implementation. Non-compliance with these criteria would – in the logic of the WWF scenario – reduce the possible share of biomass in final energy consumption and thus the share of renewables – from 41% in the fully sustainable case to only 31% with insufficient sustainable biomass available.

The authors claim to have developed the only scenario achieving a 100% target in 2050 (page 24), which need not be commented here, because it seems to be more important to put the discussion and decision process back on a road to a fully sustainable energy supply. And there is still some way to go and some debates to be successfully conducted – debates which at present are oscillating between the objective of a fully renewables based energy system and more insistingly again *"nuclear renaissance"* under the headline of striving for *"low carbon energy"* sources.

CHAPTER 10

Learning from best practice – what European legislation and policy development can contribute to global growth of renewables

Christine Lins

10.1 GLOBAL RENEWABLE ENERGY DEVELOPMENT, WHERE DO WE CURRENTLY STAND?

Renewable sources, including traditional biomass, have grown to supply 16.7%[1] of global final energy consumption in 2010. Of this total, modern renewable energy accounted for an estimated 8.2%, a share that has increased in recent years, while the share from traditional biomass has declined slightly to an estimated 8.5% (see Fig. 10.1). During 2011, renewables continued to grow strongly in all end-use sectors: power, heating and cooling, and transport.

In the power sector, renewables accounted for almost half of the estimated 208 gigawatts (GW) of electric capacity added globally during the year. Wind and solar photovoltaic (PV) accounted for almost 40% and 30% of new renewable capacity, respectively, followed by hydropower (nearly 25%). By the end of 2011, total renewable power capacity worldwide exceeded 1,360 GW, up 8% over 2010; renewables comprised more than 25% of total global power-generating capacity (estimated at 5,360 GW in 2011) and supplied an estimated 20.3% of global electricity.

In the European Union, renewables accounted for more than 71% of total electric capacity additions in 2011, with solar PV alone representing nearly half (46.7%) of new capacity that came into operation. Germany continues to lead in Europe and to be in the forefront globally, remaining among the top users of many renewable technologies for power, heating and transport. In 2011, renewables provided 12.2% of Germany's final energy consumption in 2011, 20% of electricity consumption (up from 11.6% in 2006), 10.4% of heating demand (up from 6.2%), and 5.6% of transport fuel (excluding air traffic).

The heating and cooling sector offers an immense yet untapped potential for renewable energy deployment. Heat from biomass, solar and geothermal sources already represents a significant portion of the energy derived from renewables and is slowly evolving as countries (particularly in the European Union) are starting to enact supporting policies and to track the share of heat derived from renewable sources. Trends in the heating (and cooling) sector include an increase in system size, expanding use of combined heat and power (CHP), providing renewable heating and cooling into district networks, and the use of renewable heat for industrial purposes.

Renewable energy is used in the transport sector in the form of gaseous and liquid biofuels; liquid biofuels provided about 3% of global road transport fuels in 2011, more than any other renewable energy source in the transport sector. Electricity powers trains, subways, and a small but growing number of passenger cars and motorised cycles, and there are limited but increasing initiatives to link electric transport vehicles with renewable energy.

Solar PV grew the fastest of all renewable technologies during the period from end-2006 through 2011, with operating capacity increasing by an average of 58% annually, followed by concentrating solar power (CSP), which increased almost 37% annually over this period from a small base, and wind power (26%). Demand is also growing rapidly for solar thermal heat systems, geothermal

---

[1]The statistics and graphs in this text are taken from the REN21 Renewables 2012 Global Status Report www.ren21.net/gsr, if not otherwise quoted.

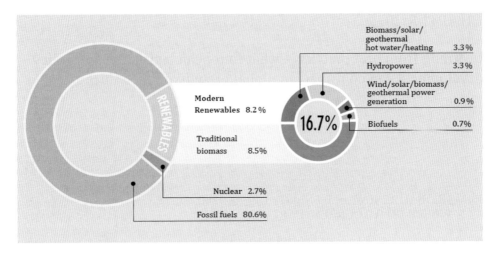

Figure 10.1.   Renewable energy share of global final energy consumption 2010.

ground-source heat pumps, and some solid biomass fuels, such as wood pellets. The development of liquid biofuels has been mixed in recent years, with biodiesel production expanding in 2011 and ethanol production stable or down slightly compared with 2010. Hydropower and geothermal power are growing globally at rates averaging 2–3% per year. In several countries, however, the growth in these and other renewable technologies far exceeds the global average.

Renewable technologies are expanding into new markets, in 2011, around 50 countries installed wind power capacity, and solar PV capacity is moving rapidly into new regions and countries. Interest in geothermal power has taken hold in East Africa's Rift Valley and elsewhere, and solar hot water collectors are used by more than 200 million households, as well as in many public and commercial buildings the world over. Interest in geothermal heating and cooling is on the rise in countries around the world, and the use of modern biomass for energy purposes is expanding in all regions of the globe.

Across most technologies, renewable energy industries saw continued growth in equipment manufacturing, sales, and installation during 2011. Solar PV and onshore wind power experienced dramatic price reductions resulting from declining costs due to economies of scale and technology advancements, but also due to reductions or uncertainties in policy support. At the same time, some renewable energy industries – particularly solar PV manufacturing – have been challenged by falling prices, declining policy support, the international financial crisis, and tensions in international trade.

10.2   STABLE POLICY FRAMEWORKS: THE ENABLING FACTOR OF RENEWABLE ENERGY DEPLOYMENT

All around the world, renewable energy targets and support policies continue to be a driving force behind increasing markets for renewable energy. Clearly, the European Union and its Member States have played a leading role in setting such targets and policy frameworks. The successful European example was replicated in many countries all around the world.

The number of official renewable energy targets and policies in place to support investments in renewable energy continued to increase in 2011 and early 2012, but at a slower adoption rate relative to previous years (see Fig. 10.2). At least 118 countries, more than half of which are developing countries, had renewable energy targets in place by early 2012, up from 109 as of early 2010.

Figure 10.2.   Policy maps 2012. The number of countries with renewable targets more than doubled between 2005 and 2012. A large number of city and local governments are also promoting renewable energy.

Several countries undertook significant policy overhauls that have resulted in reduced support; some changes were intended to improve existing instruments and achieve more targeted results as renewable energy technologies mature, while others were part of the trend towards austerity measures.

Renewable power generation policies remain the most common type of support policy; at least 109 countries had some type of renewable power policy by early 2012, up from the 96 countries reported in the GSR 2011. Feed-in-tariffs (FIT) and Renewable Portfolio Standards (RPS) are the most commonly used policies in this sector. FIT policies were in place in at least 65 countries and 27 states by early 2012. While a number of new FITs were enacted, most related policy activities involved revisions to existing laws, at times under controversy and involving legal disputes. Quotas or RPS were in use in 19 countries and at least 54 other jurisdictions, with two new countries having enacted such policies in 2011 and early 2012.

There are many variations in feed-in tariff design. Levels of support provided under FITs also vary widely and are affected by technology cost, resource availability, and installation size and

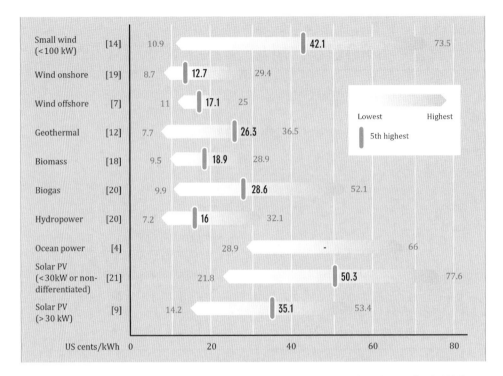

Figure 10.3.   FIT Payments for a range of renewable energy technologies, selected countries 2011/12.
Note: Each bar depicts the range of tariffs provided by selected countries with at least 15-year
terms. The number of countries analyzed by the data source is shown in parentheses; the
vertical line depicts the fifth highest tariff provided. Tariff-rates for small-scale wind, onshore
wind, geothermal, biogas and small- and large scale solar PV were updated as of May 2012;
offshore wind, biomass, and hydropower rates were updated as of September 2011; ocean
energy rates have no given year.

type (e.g., ground- versus rooftop-mounted solar PV systems). Tariffs tend to be concentrated
towards the lower end of the range for the more mature technologies of wind, geothermal, and
hydropower (see Fig. 10.3).

Historically, the highest tariffs have been set for solar PV systems of less than 30 kW of
capacity due to their relatively higher capital costs per kW, but the gap is narrowing as solar PV
manufacturing costs and market prices decline. In addition, large-scale renewable power systems
typically require lower FIT rates to be cost competitive, because they benefit from economies of
scale. Some countries have further varied rates within regional boundaries based on local resource
potential. Differentiated payments have been considered necessary for a properly functioning tariff
system. Periodic review and re-setting of rates, in line with technology and market developments,
are viewed as necessary steps for successful FIT policy implementation over time. The German
feed-in-tariff that was first introduced in 1991 has served as a model for many countries.

According to Miguel Mendonca (Mendonca 2007), *"a feed-in tariff is a renewable energy law
that obliges energy suppliers to buy electricity produced from renewable resources at a fixed price,
usually over a fixed period – even from householders. These legal guarantees ensure investment
security, and the support of all viable renewable energy technologies."*

Feed-in tariff policies can be considered as the most successful renewable energy market
introduction policies. Lately there is increasing debate about the need of advanced feed-in tariff
design once renewable energy technologies are effectively introduced into a market and reach
take-off.

Table 10.1.   Estimated jobs in renewable energy worldwide, by industry.

| Technologies | Thousand jobs | | | | | | | | |
	Global	China	India	Brazil	USA	EU	Germany	Spain	Others
Biomass	**750**	266	58		152	273	51	14	2
Biofuels	**1,500**			889	47–160	151	23	2	194
Biogas	**230**	90	85			53	51	1.4	
Geothermal	**90**				10	53	14	0.6	
Hydropower (small)	**40**		12		8	16	7	1.6	1
Solar PV	**820**	300	112		82	268	111	28	60
CSP	**40**				9		2	24	
Solar heating/cooling	**900**	800	41		9	50	12	10	1
Wind power	**670**	150	42	14	75	253	101	55	33
TOTAL	**5,000**	**1,606**	**350**	**889**	**392–505**	**1,117**	**372**	**137**	**291**

Policies to promote renewable heating and cooling continue to be enacted less aggressively than those in other sectors, but their use has expanded in recent years. By early 2012, at least 19 countries had specific renewable heating/cooling targets in place and at least 17 countries and states had obligations/mandates to promote renewable heat. Numerous local governments also support renewable heating systems through building codes and other measures.

Biofuels blending and fuel share mandates existed in at least 46 countries at the national level and in 26 states and provinces by early 2012, with three countries enacting new mandates during 2011 and at least six increasing existing mandates. Transport fuel-tax exemptions and biofuel production subsidies also existed in at least 19 countries.

Thousands of cities and local governments around the world also have active policies, plans, or targets for renewable energy and climate mitigation. Almost two-thirds of the world's largest cities had adopted climate change action plans by the end of 2011, with more than half of them planning to increase their uptake of renewable energy. Many of the institutions encouraging co-operation among cities in local renewable energy deployment saw increased membership and activities in 2011, including the EU Covenant of Mayors (with over 4,200 member cities). The Covenant of Mayors[2] is the mainstream European movement involving local and regional authorities, voluntarily committing to increasing energy efficiency and use of renewable energy sources on their territories. By their commitment, Covenant signatories aim to meet and exceed the European Union's 20% CO_2 reduction objective by 2020.

Policymakers are increasingly aware of renewable energy's wide range of benefits – including energy security, reduced import dependency, reduction of greenhouse gas (GHG) emissions, prevention of biodiversity loss, improved health, job creation, rural development, and energy access – leading to closer integration in some countries of renewable energy with policies in other economic sectors. Globally there are more than 5 million jobs in renewable energy industries, and the potential for job creation continues to be a main driver for renewable energy policies. During 2011, policy development and implementation were also stimulated in some countries by the Fukushima nuclear catastrophe in Japan and by the UN Secretary-General's announced goal to double the share of renewables in the energy mix by 2030.

The European experience also shows that the countries which put in place stable policy frameworks are the ones benefitting most from jobs created in the renewable energy sector. Out of approx. 550.000 persons employed in the renewable energy industry in Europe, about 387.000 jobs are located in Germany.

---

[2] www.eumayors.eu

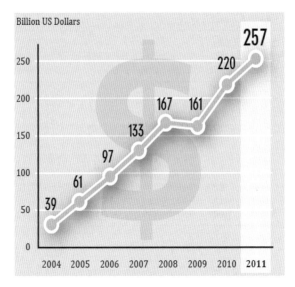

Figure 10.4.   Global trends in renewable energy investment 2011 (Source: UNEP/Bloomberg, Graph GSR 2012).

10.3   MONEY FLOWS WHERE POLICY STABILITY IS PROVIDED

Global new investment in renewables rose 17% to a record USD 257 billion in 2011 (see Fig. 10.4). This was more than six times the figure for 2004 and almost twice the total investment in 2007, the last year before the acute phase of the recent global financial crisis. This increase took place at a time when the cost of renewable power equipment was falling rapidly and when there was uncertainty over economic growth and policy priorities in developed countries. Including large hydropower, net investment in renewable power capacity was some USD 40 billion higher than net investment in fossil fuel capacity.

One of the highlights of 2011 was the strong performance of solar power, which blew past wind power, the biggest single sector for investment in recent years. The top five countries for total investment were China, which led the world for the third year running, followed by the United States, Germany, Italy, and India. India displayed the fastest expansion in investment of any of the large renewables market in the world, with 62% growth.

10.4   GLOBAL TARGET SETTING

The United Nations General Assembly declared 2012 the International Year of Sustainable Energy for All. UN Secretary-General Ban Ki-moon has supported the Year with his new global initiative – Sustainable Energy for All, which seeks to mobilize global action on three inter-linked targets to be achieved by 2030: universal access to modern energy services, improved rates of energy efficiency, and expanded use of renewable energy sources.

On the one hand, it is interesting to note the trend in the UN approach to link demand and supply side. So far, there has been little systematic linking of energy efficiency and renewable energy in the policy arena, but countries are beginning to wake up to the importance of tapping their potential synergies. Efficiency and renewables can be considered the *"twin pillars"* of a sustainable energy future, with renewables reducing the emissions of pollutants per unit of energy produced, and energy efficiency improvements reducing energy consumption altogether. Improving the efficiency of energy services is advantageous irrespective of the primary energy source, but there is a special synergy between energy efficiency and renewable energy sources.

On the other hand, the UN Secretary General's initiative is somehow comparable to the approach the European Union has taken with setting indicative targets as outlined in its 1997 White Paper on Renewable Energy (European Commission 1997c). It goes without saying, however, that a global move from indicative to binding targets for renewable energy will be more challenging in view of the complex international governance structure.

However, the framework of target setting for 2030 could be somehow inspiring for the European institutions where the setting of a 2030 target is currently debated and where the European renewable energy industry has underlined the need for such a target (EREC 2011).

10.5   FUTURE OUTLOOK

Renewable energy policy frameworks, originally introduced in high-income countries, are spreading to more and more countries all around the world. Renewable energy policies are put in place to address questions of energy security, climate change, economic development and job creation.

Beyond the binding 2020 renewable energy targets, different European countries are currently debating the role of renewable energy sources in their future energy mix. In France, for example, newly elected President Francois Hollande has announced to reduce the share of electricity from nuclear from 75 to 50% by 2025. In order to reach this, a nation-wide debate between public and private sector stakeholders was initiated. For the first time in history, the composition of the future energy mix of France is publicly debated; so far this has been solely decided by the political elite.

In Germany, for example, the Fukushima nuclear accident in March 2011 led to a more rapid exit from nuclear energy use, now scheduled for 2022. To this end, a decision was taken to completely reform the nation's energy sector through what is referred to as "Energiewende" (Energy Transition), which focuses on energy efficiency and renewable energy sources together with massive energy infrastructure investments. The "Energiewende" is Germany's biggest modernization and infrastructure project that will contribute to energy security, jobs and value creation. Due to Germany's status as one of the world's leading economies, the energy transition project has frontrunner character with a global impact.

Several scenarios including from the International Energy Agency and others predict a high share of renewable energy globally by 2050. The IPCC's Special Report on Renewable Energy Sources and Climate Change Mitigation (IPCC 2012) indicates that close to 80% of the world's energy supply could be met by renewables by mid-century if backed by the right enabling public policies.

With its Climate and Energy policy, the European Union has been a front-runner in international renewable energy policy design, at least for the horizon 2020. Current discussions on a renewable energy policy framework until 2030/2050 are ongoing; their outcome will show whether the European Union manages to keep its pole position in international renewable energy promotion.

CHAPTER 11

Towards an integrated post-2020 framework

Rainer Hinrichs-Rahlwes

11.1 LOOKING TOWARDS 2030

The scene is set. The 2020-framework for the European Union is in place – expiring, however, in 2020, if no further agreement is reached in due time. And a rough outline of an international climate roadmap was developed in December 2012 in Durban (South Africa), which may or may not result in a new climate regime in 2015. Even with a new agreement reached in 2015, there will most likely not be any significant additional incentives for shifting supply and demand to a fully sustainable, renewables based energy system. The EU (and other major economies) will have to take responsible decisions about a future framework for climate and energy policies, which could serve as incentives for the 2015 negotiations – but success cannot be taken for granted. Own efforts to move towards a renewables based energy supply would definitely strengthen the EU's contribution to reducing global greenhouse gas emissions. Sustainable energy policies for Europe would also provide significant benefits regarding clean environment and economic growth, security of energy supply and affordable energy for all, job creation and competitiveness. Those Member States, which most efficiently and successfully increase their shares of renewable energy will benefit from their frontrunner role.

This chapter analyses the ongoing discussion in the EU about the post-2020 climate and energy framework, where renewable energies and energy efficiency as well as energy saving should be in the centre. The first and foremost beneficiary of such a strategy would be global climate protection, because this would deliver on greenhouse gas emissions reduction in an efficient and cost effective way.

The European Commission's communication *"Renewable Energy: a major player in the European energy market"* (European Commission 2012b) including accompanying impact assessment (European Commission 2012c, European Commission 2012d) and Commission Staff Working Document (European Commission 2012e) will be analysed, as well as the process towards *"State Aid Modernisation (SAM)"* launched in parallel by a Commission's Communication in May 2012 (European Commission 2012a). In addition, the Commission's Proposal for amending the Renewables Directive and the Fuel Quality Directive (European Commission 2012f) including accompanying documents (European Commission 2012g, European Commission 2012h) regarding ILUC (Indirect Land Use Change) will be examined with regard to potential impact on achieving 2020-targets and on the development of a post-2020 framework.

Discussions and developments focusing on the traditional energy sector are presented, because decisions regarding support for nuclear and CCS as well as the future of the EU-ETS as Europe's main instrument for carbon pricing will have a major impact on the economic viability of sustainable energy policies for renewables and efficiency. A closer look is therefore taken on the Commission's *"Backloading"* proposal (European Commission 2012i), including accompanying Impact Assessment (European Commission 2012j, European Commission 2012k), an attempt to stop the downwards trend of carbon prices by temporarily setting aside excess allowances.

Eventually, the European Parliament's discussion on a post-2020 framework as laid out in a report initiated by the ITRE-Committee (European Parliament 2012, European Parliament 2013) is analysed, as well the European Commission's Green Paper *"A 2030 framework for climate and energy policies"* (European Commission 2013a) and the *"Renewable energy progress report"* (European Commission 2013c). The documents will be analysed and assessed regarding triggers and hurdles they may provide for the EU to move forward towards an ambitious and integrated climate and energy framework for the next decade, including 2030-targets for renewable energy and energy efficiency.

11.2   RENEWABLES – A MAJOR PLAYER IN THE ENERGY MARKET

In December 2011, the European Commission opened a "Public consultation on the Renewable Energy Strategy" (European Commission 2011m), intending to facilitate the discussion on a post-2020 framework for renewable energy policies. A few months after the 2020 framework had entered into force and implementation had started, transposition into national law had by far not been completed and there was first evidence of hesitation and reluctance in some of the Member States. In May 2011, the European Renewable Energy Council (EREC) had launched a campaign asking for a binding renewables target of 45% in 2030 (EREC 2011): *"EREC calls on the European Commission, Member States and the European Parliament to deliver on the European Union's long-term climate commitment by proposing and endorsing a legally binding EU target of at least 45% renewable energy by 2030"*. Other stakeholders and NGOs had started asking to prepare agreement on the next milestone after 2020, which is looking at 2030.

When the public consultation was launched, the consultation document (European Commission 2011m) addressed all major questions which need to be answered for an informed decision to be reasonably taken. The Commission assessed that a process should be initiated in due time, although the Renewables Directive calls for a post-2020 Roadmap only in 2018. Highlighting multiple benefits renewables have provided so far the document explores various approaches to setting new targets and milestones. *"It is therefore only appropriate to consider establishing standalone targets for renewable energy sources post-2020 in the context of possible post-2020 targets for energy efficiency and climate mitigation including the functioning of the ETS"*. The consultation focuses on the question *"to what extent and in which form policy intervention on EU and national levels will continue to be needed after 2020"* (page 2). In particular, incentive schemes, network issues, regional and interregional cooperation, technology development and sustainability are addressed. Among others, the role of targets is explored, as well as whether and to what extent financial support is going to be needed after 2020 and whether this should be on a national and/or a European level. Administrative procedures are assessed, as well as quantity and quality of infrastructure needs. Specific attention is paid to market integration, exploring how – when after 2020 *"renewables will represent a significant share of the market"* (page 6) – they could be *"made responsive to market signals"* and how it can *"be ensured that market arrangements reward flexibility"*, which is needed for the future energy system. There are questions about heating and cooling, too, and about transport fuels and sustainability, as well as questions about regional and interregional dimensions.

The findings from the consultation were presented to the public on 24 February 2012 in Brussels[1]. A major outcome was a broad majority asking for mandatory 2030-targets for renewables, some more for indicative 2030-targets; there was also significant support for a mix of national und EU targets. As a result of the consultation, the existing mechanisms and regulations need to be strengthened and improved.

Four months later, the Commission presented an evaluation of the consultation in the form of a strategy paper: *"Renewable Energy: a major player in the European energy market"* (European Commission 2012b). As an introduction, the benefits of renewables are highlighted. *"Renewable energy enables us to diversify our energy supply. This increases our security of supply and improves European competitiveness creating new industries, jobs, economic growth and export*

---

[1] http://ec.europa. eu/energy/renewables/events/doc/20120224_res_presentations.zip

opportunities, whilst also reducing our greenhouse gas emissions. Strong renewables growth to 2030 could generate over 3 million jobs, including in small and medium sized enterprises. Maintaining Europe's leadership in renewable energy will also increase our global competitiveness, as 'clean tech' industries become increasingly important around the world [. . .] The renewable energy goal is a headline target of the Europe 2020 strategy for smart, sustainable and inclusive growth" (page 2).

The need for stable framework conditions to secure and continue the benefits is underlined: *"In parallel to a rigorous implementation and enforcement of the Renewable Energy Directive, clarity on longer term policy is needed to ensure that the necessary investment is made"* (page 2). Based on the analysis of the Energy Roadmap 2050 (European Commission 2011g) that in all decarbonisation scenarios *"Regardless of scenario choice, the biggest share of energy supply in 2050 will come from renewable energy"* and *"Strong growth in renewables is the so-called 'no regrets' option"* (page 3) the post-2020 strategy is developed.

The Commission finds that significant *"maturing"* of some renewables technologies is being observed and costs have decreased considerably, particularly in solar PV, wind onshore and some biomass technologies. They find that *"Europe's renewable energy sector has developed much faster than foreseen at the time of drafting the Directive"*. Therefore it needs to be ensured that *"renewable energy technologies become competitive and ultimately market driven"* (page 4). However, to facilitate this process, *"Policies which hinder investment in renewables should be revised and in particular, fossil fuel subsidies should be phased out"*. Based on these recommendations, *"renewable energy should be gradually integrated into the market with reduced or no support, and should over time contribute to the stability and security of the grid on a level footing with conventional electricity generators and competitive electricity prices. In the longer term, a level playing field needs to be ensured"*. On a level playing with subsidies for fossil or nuclear energy no longer creating competitive advantages for unsustainable energy most renewables will sooner or later be able to compete freely. However, there is a long way to go until this will be achieved. The Commission's statement that exposure to market risks is urgent, whereas the level playing field needs to be ensured *"in the longer term"* certainly is problematic priority setting, which might put renewables development at risk. And of course, the statement that no more support will be needed does not fully take into account that there are (and will be for a long time) some technologies which are under development and therefore need specific support for market introduction to achieve maturity and necessary scale.

The balance between market integration and remaining need for support (and R&D) is a main element of the strategy paper, where – in some sentences – it becomes clear that the Commission might be pushing too fast for *"market integration of renewables"* instead of pushing for completing the internal energy market and making it work. This can be seen e.g. in the following statement: *"Moving as rapidly as possible towards schemes that expose producers to market price risk encourages technology competitiveness"*, followed by another major misperception, now with regard to national support schemes. *"Moreover diverging national support schemes, based on differing incentives may create barriers to entry and prevent market operators from deploying cross-border business models, possibly hindering business development. Such a risk of impairing the single market must be avoided and more action is also needed to ensure consistency of approach across Member States, to remove distortions and develop renewable energy resources cost effectively"* (page 5). In reality, diverging national support schemes have removed diverging barriers against renewable energy in the different countries, cross border issues being the least of all concerns in this context, but rather the uncontested dominance of incumbent utilities in combination with incomplete implementation of – among various other issues – the unbundling regulations of the Internal Energy Market Packages (see Chapter 7 of this book).

The strategy paper addresses another important barriers preventing renewables from becoming equal players on a level playing field: the dominant market design based on conventional (inflexible and baseload driven) power systems. It is correctly outlined that the role of multiple power producers and consumers needs to be taken into account. In particular *"wholesale electricity prices, based on short run marginal costs, may face downward pressure due to the rise* of wind and solar power (with near zero marginal costs). The market should be able to respond, reducing supply when prices are low and increasing it when prices are high. Changes in market prices need to encourage flexibility, including storage facilities, flexible generation, demand-side management (as consumers respond to changing price patterns)" (page 7). It is underlined that capacity payments alone cannot solve this problem (even less so, if not well designed). "Such an approach may encourage investment, but it also separates investment decisions from market price signals. Moreover, if poorly designed, it could "lock in" solutions focused on generation that frustrate the introduction of new forms of flexibility. Aggregated distributed generation, demand response and expanded balancing areas would also be impaired" (page 7). The necessary debate about appropriate market design is opened.

The need for adapting infrastructure is addressed: transmission grids and even more urgently distribution networks need to be enhanced and enabled to cope with increasing shares of flexible and more distributed power production and demand. This is seen as a crucial condition for developing the single energy market, and for consumers to be able to choose their suppliers Europe-wide and thus create competition and drive prices down.

The issue of sustainable biofuels and (as mid-term perspective) all biomass is addressed by announcing more analysis and – if appropriate – proposals for including and enforcing sustainability criteria of the Renewables Directive, including assessment about whether and how to integrate this aspect in a post-2020 framework.

Finally, the strategy paper presents recommendations regarding further development of the policy framework. Whereas the existing overall framework is considered "to work well" (page 12), further development needs to be facilitated and guided. The Commission foresees to present guidance about good support schemes and probably about support scheme reform, as well as guidance on application of the cooperation mechanisms of the directive[2]. In order to provide guidance on further development of a stable and reliable framework for renewables post-2020, three options (apart from a business as usual case) for the future of support schemes are described and analysed in the accompanying impact assessment (European Commission 2012e). This is also seen as important precondition for assessing "concrete 2030 milestones" (page 12).

The options are further described (page 6):

(1) "Business as usual [. . .],
(2) Decarbonisation without renewable energy targets post-2020 [. . .],
(3) Binding renewable energy targets post-2020 and coordinated support [. . .],
(4) EU renewable energy target and harmonized measures [. . .]".

The Communication underlines that the business-as-usual scenario is not an option to deliver on sufficient further growth and development, neither of greenhouse gas reduction nor of renewables. Although not explicitly stated in the strategy paper, it is clear from the impact assessment that option (3) is the one which deserves most support – in contrast to options (2) and (4), all of which would not be able to deliver as precisely and successfully on renewables growth and related benefits. The impact assessment is clear when it comes to the advantages of option (3): "Binding renewable energy targets post-2020 and coordinated support. Depending on their ambition, targets could help provide investors and the business community with greater certainty on future market volumes for renewable energy technologies. They would also promote further cost-effectiveness and convergence of national support schemes and foster greater research and development of innovative technologies. This option would also effectively address sustainability and public acceptance issues, by promoting a more balanced and regionally equilibrated deployment of renewables" (page 6). Renewable energy industry and associations have supported this assessment and are arguing for focussing on this option for further development of a supportive policy framework.

----

[2]When this book was finished, publication of the guidance papers seemed likely in July 2013.

Like all Commission's communications, the Renewable Energy Strategy was debated and discussed in the Council and in the European Parliament. Whereas the European Parliament took some time to analyse the strategy and eventually decided to integrate comments to the strategy into a special report on *"Current challenges and opportunities for renewable energy on the European energy market"* (European Parliament 2012, European Parliament 2013 – see later in this chapter), the Council agreed on conclusions in December 2012 (Council 2012). In the conclusions, the achievement of the Renewables Directive in *"providing certainty to investors and a stimulus up to 2020"* is highlighted, as well as the expectation *"that a longer-term perspective would have a positive influence on investments, given the long planning process and investment horizon for renewable energy sources, while ensuring economic and environmental sustainability* (page 1). Given the need for a mid-term perspective, the Council welcomes the Communication, *"which provides useful perspectives for initiating the reflection on a consistent post-2020 framework to maintain a robust growth of renewable energy"* (page 2). The completion of the internal energy market by 2014, as foreseen, is a major priority to be pursued, as well as market integration of renewable energy and continuous improvement of national support schemes. The Council welcomes the Commission's intention to develop non-binding guidance *"on support scheme reform"* (page 3) and *"on the implementation of the cooperation mechanisms"* (page 5). The Council also highlights the *"need for the rationalization and the phasing out of environmentally or economically harmful subsidies, including for fossil fuels"* (page 3). The need for addressing infrastructure issues, consumer protection, innovation and sustainability is underlined. Eventually, the *"Council invites the Commission to present in appropriate time and after thorough analysis, discussion and the review by 2014 of certain aspects of the current Renewable Energy Directive as foreseen therein, a solid and effective EU post-2020 RES framework embedded in the broader context of and contributing to the long-term overall EU policy framework. The RES post-2020 framework should take into account, inter alia, the experience gained with the current RES policy framework, including its cost-effectiveness and the interactions between different targets and instruments, and its implementation"* (page 6). This statement was welcomed by the renewable energy industry as setting *"the right conditions for the Commission to now start work on an ambitious 2030 renewables target"* (EREC 2012).

11.3   RENEWABLES AND STATE AID REGULATIONS

Creating an internal market and preventing it from distortion *"by anticompetitive behaviour of companies or by Member States favouring some actors to the detriment of others"* (European Commission 2012a, page 2) is one of the main objectives of the EU. Therefore all public spending needs to comply with this overarching consensus[3]. Some support schemes for renewable energy, which are using public funds or public budgets for investment and/or operational aid are subject to the European Union's State aid regime. They have to be notified to the European Commission for scrutiny whether or not the support system (including the details of payment to energy producers) is necessary and proportionate for achieving objectives within the consensus of the EU Treaties and which cannot reasonably be achieved otherwise. Protection of market players against powerful incumbents is a possible justification for State aid. Support for renewable energy is basically considered to be in compliance with State aid regulations and therefore included in the Environmental Aid Guidelines, because they are considered to be a relevant means for greenhouse gas mitigation and other environmental beneficial objectives. Nevertheless, although support for renewable energy is a legitimate aim justifying State aid, it is still necessary to notify the details of planned regulations so that the European Commission can assess the proportionality of the foreseen tariffs and regulations.

---

[3]A summary of State Aid rules in the EU can be found on the European Commission's Website: http://europa.eu/legislation_summaries/competition/state_aid/, viewed 6 May 2013.

Apart from those support mechanisms, which are State aid and which are therefore treated as such by the Commission (e.g. Austria, Luxembourg, Slovenia), there are other national support systems, which do not fall under the State aid regime, because they are not fed by public funds or from the public budget. The most outstanding of these systems is the German feed-in legislation as set up in 1991 and refined in the *"Renewable Energy Sources Act"* (EEG) in 2000 (and in several amendments since then)[4]. The question, whether the EEG (and the predecessor law of 1991) is State aid or not, has been discussed for more than a decade. Until today, there is an overwhelming consensus among legal experts that the ruling of the European Court of Justice of 13 March 2001 (ECJ 2001) is still applicable to the EEG in its present version. However, when the text of this book was finalized, at least one legal case was still pending, challenging the EEG as unlawful State aid. According to some media, the European Commission (Directorate General Competition) had started a procedure to assess whether the present version of the law has to be considered as State aid, and if so whether this is lawful or unlawful. It appears to be obvious that the EEG cannot to be classified as State aid, at least not as unlawful[5], but that will only be proven after the proceedings on this case are closed.

State aid is not only important for sustainable energy policies in individual law cases, but further development of the EU State aid framework will have a strong impact on the future of renewable energy and greenhouse gas reduction policies. The European Commission has started a process of *"EU State Aid Modernisation (SAM)"* (European Commission 2012a), aiming at simplification and streamlining and more focussed application of the State aid regime. *"In particular, public spending should become more efficient, effective and targeted at growth-promoting policies that fulfil common European objectives"* (page 2). State aid, if necessary, should contribute to *"ending the crisis and re-igniting sustainable growth"*. And it should be *"well-designed, targeted at identified market failures and objectives of common interest, and least distortive ('good aid')"* (page 4). *"Good aid"* would enforce the *"polluter pays principle"* and it would *"encourage companies to go beyond mandatory EU environmental standards or to promote energy efficiency provided for in the Environmental aid guidelines"* (page 4–5). The European Commission is calling for *"Robust State aid control [...] to ensure a well functioning single market"*. This is important and potentially beneficial for renewable energy and efficiency development, as the Commission finds robust control to be *"of particular relevance in markets that have only recently been opened and where large incumbents aided by the State still play a major role, such as transport, postal services or, in more limited cases, energy. State aid modernisation can improve the functioning of the internal market through a more effective policy aimed at limiting distortions of competition, preserving a level playing field and combating protectionism"* (page 5). Properly applied, this could limit market dominance of incumbents and thus pave the way for new market entrants to fairly compete on a level playing field. It is foreseen to complete the process of SAM, including *"revision and streamlining of the main Commission acts and guidelines by the end of 2013"* (page 9).

Based on a public consultation and consultation of Member States new regulations and guidelines will be developed. For renewable energy industry and investors, it will be of particular importance that consistency of regulations regarding support schemes is maintained and constructively developed. Inclusion of renewable energy in the Environmental aid guidelines has proven to be useful for handling those support schemes which are subject to the State aid regime. Therefore, as long as markets remain distorted due to dominant incumbents and lack of internalization of external costs for fossil and nuclear fuels, this needs to be maintained or replaced by an even more comprehensive framework. And the opportunity should be taken to clarify that

---

[4]The latest version in English is on the Website of the German Ministry for the Environment: http://www.bmu. de/en/service/publications/downloads/details/artikel/renewable-energy- sources-act-eeg-2012/

[5]An overview of recent developments and legal assessment – only in German so far – at *Stiftung Umweltenergierecht* (translated: Foundation Environmental Energy Law) http://www.stiftung-umweltenergierecht. de/fileadmin/pdf/wiss._Veroeff/Stiftung_Umweltenergierecht_Beihilfediskussion_EEG_2013-03-11.pdf

fossil and nuclear energy are not eligible for lawful State aid. Regarding fossil fuel, the European Commission and others have repeatedly underlined the need to phase out subsidies (although there are some back-doors for CCS). For nuclear, however, there is a strange mix of not mentioning the problem on the one side and trying to include support for nuclear as *"low carbon energy"* among eligible technologies for lawful State aid[6]. This has to be clearly rejected in the ongoing discussion process, if Europe is to continue towards a truly sustainable energy system.

On the other hand, it needs to be clarified in the modernized framework that State aid for renewable energy is allowed until market failures, which prevent renewables from fairly competing on a level playing field, are removed. As such, renewables support is not distorting competition but facilitating fair competition in distorted markets. Creating block-exemption rules for renewables support could be a positive step, if carefully designed and actually removing competitive advantages of fossil and nuclear energy.

11.4   THE FUTURE OF BIOFUELS AND BIOMASS – IN TRANSPORT AND BEYOND

Sustainability of biofuels (and biomass in general) is an important issue related to transforming the energy system towards renewable energy. In the Renewables Directive 2009 and in the Fuel Quality Directive 2009 sustainability is addressed by a number of regulations and review clauses. For biofuels and liquid biomass sustainability criteria are included and reporting obligations for the Commission invite additional proposals for all biomass and for improving the sustainability criteria for liquid fuels. Whereas the discussion and decision about extending sustainability criteria to solid biomass is pending and legislative proposals are not expected in the near future, the Commission has taken the initiative (European Commission 2012f) to propose amendments to both directives regarding indirect land use change (ILUC). The proposal which is controversially discussed in Council working groups and in the European Parliament's Committees aims at limiting support and accountability of food based biofuels *("conventional biofuels")* and instead incentivising *"advanced biofuels"*. Given the fact that biofuels will have to provide the major contribution to the 10% renewables-in-transport target, the proposal suggests several measures to facilitate market penetration of *"advanced biofuels"*. Key instrument for achieving this objective is the inclusion of ILUC as a major element for sustainability assessment of biofuels – despite ongoing scientific and political discussions about the validity of existing modelling exercises and methodologies, thus moving away from addressing ILUC *"under a precautionary approach"* (page 2). The proposal intends to *"limit the contribution that conventional biofuels (with a risk of ILUC emissions) make towards attainment of the targets in the Renewable Energy Directive"* to a maximum of 5% and to *"improve the greenhouse gas performance of biofuel production processes (reducing associated emissions) by raising the greenhouse gas saving threshold for new installations"* to at least 60% above conventional fuels as of 1 July 2014. Whilst protecting existing installations these changes are deemed necessary, as well as improved reporting requirements for Member States and particularly encouraging *"a greater market penetration of advanced (low-ILUC) biofuels by allowing such fuels to contribute more to the targets in the Renewable Energy Directive than conventional biofuels"* Furthermore, it is suggested to phase out support for conventional biofuels after 2020.

---

[6]A most recent case is a *"Consultation Paper"* on *"Environmental and Energy Aid Guidelines 2014–2020"* (European Commission 2013) of March 2013. Whilst highlighting the need to phase out support for *"the most mature RES"* (page 14), the paper asks for *"increased competition between low-carbon sources"* (page 8) and at the same time suggests *"in-depth discussion in order to analyse whether market failures justify intervention"* in order to *"widen support also to other low-carbon energy sources including nuclear merits"* (page 13). Although this is absurd, there are some Member States seriously asking for State aid for nuclear – after 70 years of subsidies.

In order not to risk the achievement of the 10% renewables-in-transport target for 2020, it is proposed to numerically raise the contribution of *"advanced biofuels"* by *"increasing the weighting of advanced biofuels towards 10% target for transport set in Directive 2009/28/EC compared to conventional biofuels"* (page 7) – counting those fuels up to fourfold (page 14). Compliance with the 10% target could thus be achieved by 5% *"conventional biofuels"* and only 1.25% of *"advanced biofuels"*. This proposal has been criticised for undermining the credibility of the biofuels target in a first step and the 20% renewables overall target as well. Before December 2017, the Commission shall submit a report *"reviewing [. . .] the effectiveness of the measures introduced by this Directive in limiting indirect land-use change greenhouse gas emissions associated with the production of biofuel and bioliquids. The report shall, if appropriate, be accompanied by a legislative proposal [. . .] for introducing estimated indirect land use change emissions factors into the appropriate sustainability criteria to be applied from 1st January 2021 and a review of the effectiveness of the incentives provided for biofuels from non-land using feedstocks and non-food crops [. . .]"* (page 17).

Whereas some environmental NGOs have welcomed the proposal for trying to limit the use of biofuels, industry and agricultural stakeholders have criticised it for wilfully applying a methodology which will not deliver on limiting direct or indirect land use change, but which will further reduce economic viability of alternatives to conventional transport fuels. It is emphasized that generalizing ILUC factors will not have a relevant impact on land use change in the three major biofuel producing and exporting countries (Malaysia, Indonesia, Brazil), but it will primarily have a negative impact on sustainable European biofuels production. The discussion about the Commission's proposal and alternative suggestion – such as using regional and/or country specific criteria instead of generalizing factors – is ongoing. When this book was written, it could not yet be predicted, if and when and with which modifications the Commission's proposal will be adopted by Council and European Parliament.

Given the crucial importance of biofuels for replacing oil in the transport sector and given the slow development of *"advanced biofuels"* as well as electromobility, the proposal is immature (due to lack of reliable evidence and derived methodology) and detrimental for further progress towards sustainable transport fuels. Given the fact that domestic or regional biofuels production very often does not raise (strong) sustainability concerns but provides viable solutions for decarbonising the transport sector, it seems to be short sighted and lacking vision to impose a cap on conventional biofuels and to mainly rely on *advanced* fuels (and on electromobility – but this is not addressed in the ILUC-proposal). It would therefore be appropriate to basically dismiss the ILUC proposal and replace it by regulations encouraging further development of sustainable biofuels, combined with further incentives for more advanced biofuels as well as strong and effective support for take-off of electromobility.

11.5   EMISSIONS TRADING: TRYING TO REPAIR A KEY SYSTEM

The European Emissions Trading System was designed to become a major instrument for decarbonisation. Some supporters had even argued (and still are arguing) it should become the only instrument for incentivising greenhouse gas reduction in the EU. However, in the first trading period, too many allowances had been allocated so that no relevant price signal was emerging from the system. This is why – in the Climate and Energy Package 2009 – major changes had been implemented, the most important one being the introduction of auctioning as the regular distribution method instead of grandfathering as in the first period. However, this alone did not solve the problem. Carbon prices continued to be far below the levels which are supposed to have relevant impact on investment decisions in favour of low carbon emitting sources. The low carbon prices not only undermined the effectiveness of EU-ETS for greenhouse gas reduction, but they also resulted in lower auctioning revenues – revenues which Member States had planned to use for supporting greenhouse gas reduction in non-ETS sectors.

Building on these observations, in November 2012, the European Commission presented a draft Regulation suggesting *"Backloading"* of some of the greenhouse gas allowances due for auctioning (European Commission 2012i). As usual, the proposal was accompanied by an *"Impact Assessment"* (European Commission 2012j, European Commission 2012k). It builds on the Emissions Trading Directive 2009, allowing – under certain circumstances and with the Council and the European Parliament agreeing – to change the annual amount of allowances to be auctioned within the trading period (2013–2020). Taking into account *"exceptional changes in drivers determining the balance between the demand for and supply of allowances"* the Commission underlines the necessity for immediate action. The changes addressed are *"notably the renewed economic slowdown, as well as temporary elements directly related to the transition to phase 3, including increasing unused volume of allowances valid for the second trading period for compliance in the said period, increasing volumes of certified emission reductions and emission reduction units from emission reduction projects under the Clean Development Mechanism or under Joint Implementation provisions for surrendering by operators covered by the scheme"* (European Commission 2012i, page 2). In order to reach an agreement with Council and Parliament as soon as possible, the Commission did not undertake to propose fundamental changes to the ETS-Directive, which are necessary in the mid-term, but only to limit the effects of overallocation in the *"transition period"* 2013–2015. Estimating a *"surplus of almost 1 billion allowances at the end of 2011"* (European Commission 2012k, page 2) which will continue growing, six options – in addition to a business-as-usual option – are developed and assessed with regard to their respective impact on carbon price development. All options consist of reducing the number of allowances to be auctioned in 2013, 2014 and 2015 and *"Backloading"* them in 2018, 2019 and 2020 (or only in 2020). The strongest positive impact (i.e. increasing carbon prices in the first three years) is expected from backloading 1,200 million allowances. The most limited effect derives from backloading only 400 million allowances. On the other hand, backloading 1,200 million is expected to have very negative impacts on price development at the end of the trading period, when such a high number of allowances would drastically reduce prices, when being re-introduced into the system. As a result of the impact assessment, the proposal is to *"backload"* 900 million (400 million in 2013, 300 in 2014 and 200 in 2015) and to bring them back in 2019 and 2020 (300 million in 2019, 600 in 2020) (European Commission 2012i, page 3).

Despite the limited impact of the proposal, the decision process was very difficult. Until this book was finished, an agreement on the proposal between the European Parliament and the Council had not yet been reached. Whereas the Environmental Committee of the European Parliament supported it, the Industry and Energy Committee (ITRE) did not. The European Council did not yet decide, because some countries (e.g. Poland) strictly opposed the proposal and others – e.g. Germany – had a blocked constellation in the government coalition, with the Environment Minister in favour and the Minister for Economic Affairs opposing. On 16 April 2013 the European Parliament's plenary vote followed the ITRE Committee and rejected the backloading proposal. It remains to be seen now, whether the upcoming negotiations between the European Parliament and the Council can change the rejecting vote and craft an agreement on at least some backloading – or maybe to start a process of more in depth revision.

In the meantime, the EU-ETS is seriously affected by significant overallocation and resulting prices far below any potential impact threshold. Although a timely agreement on the backloading proposal would be of paramount importance, such an agreement could only be a first *"signal of willingness/capability to act (avoids further price crash)"* (Matthes 2013, page 9). More coherent measures are needed to recover the ETS as a relevant instrument for carbon pricing and greenhouse gas reduction. Significant set-asides and further reduction of allowances will be necessary, as well as phasing-out the use of CDM- and JI-allowances from the EU-ETS. Additional flexibility needs to be established, e.g. indexation following indicators such as economic growth or development of energy consumption and/or other means of greenhouse gas reduction, such as renewable energy, energy efficiency or energy savings. Introducing minimum carbon prices or energy taxation could help revitalise the EU-ETS, to transform it into the effective instrument for greenhouse gas reduction, which it could be, if allowances were regularly and ambitiously reduced and if it were coherently integrated in a comprehensive framework of sustainable energy policies in Europe.

11.6   ANOTHER ROUND OF WISHFUL THINKING

Developing Carbon Capture and Storage (CCS) as a relevant contribution to greenhouse gas reduction was already part of the Climate and Energy Package 2009, with the CCS-Directive providing a framework for take-off. Today, virtually no progress has been made so far towards economic viability of CCS. The European Commission has therefore published a Communication *"on the Future of Carbon Capture and Storage in Europe"* (European Commission 2013d), launching a public consultation asking stakeholders about their views *"on what would be the best policy framework to ensure that demonstration and further deployment of CCS, if proven commercially and technically viable, take place without further delay"* (European Commission 2013e). The communication shows how far away CCS is from commercial scale deployment. And it demonstrates how desperately supporters are trying to find a justifying rationale and ways to develop it.

The Commission makes the case for CCS by mentioning the high shares of fossil fired power plants still in operation and under construction, a trend which is assumed to continue in the next decades. Such a trend is not in line with decarbonisation targets and with the aim of limiting global temperature increase, so that CCS technology *"is one of the key ways to reconcile the rising demand for fossil fuels with the need to reduce greenhouse gas emissions"* and *"globally CCS is likely to be a necessity to keep average global temperature rise below 2 degrees"*. It is admitted that there are other options for decarbonisation, but – referring to the Low Carbon Roadmap 2050 (European Commission 2011b) and the Energy Roadmap 2050 (European Commission 2011f) – the EU's commitment to *"supporting CCS both financially and with regulatory steps"* (page 3) is repeated, whilst conceding that so far not a single commercial size demonstration project could be built in the EU, despite ETS and additional funding programmes (e.g. the European Energy Programme for Recovery[7] and the NER300[8] Programme). The communication reiterates the commitment to reducing greenhouse gas emissions by at least 80% by 2050 and then evokes the need for CCS. *"Therefore, the 2050 target can only be achieved if the emissions from fossil fuel combustion are eliminated from the system, and here CCS may have an essential role to play, as a technology that is able to significantly reduce CO_2 emissions from the use of fossil fuels in both the power and industrial sectors"* (page 11), adding that *"it can also be used to capture biogenic carbon from the use of biomass (Bio-CCS)"* (page 12) – which is probably true, but does not help to overcome any of the real problems, like costs and public acceptance.

The communication finds that only two of 20 CCS demonstration projects are operating in Europe (in Norway, where a tax of 25 €/tCO_2 is due, and not in the EU). And both Norwegian projects are only viable in combination with enhanced oil and gas recovery. Given that the costs per tonne of avoided CO_2 in current demonstration projects are in the range of 40 € for coal and 80 € for gas, this is far from economic viability, even with support from public spending, and even further away, when carbon prices remain in the range of 5 €/t CO_2. The communication rightly analyses that *"the business case significantly worsened as of 2009"* (page 16). Anyway, it finds that *"the first generation CCS coal or natural gas power plants are expected to be significantly more expensive than similar conventional plants without CCS"*. And as if this were just a minor problem, faith healing can cure it. *"Once CCS power plants start being deployed, costs will decrease benefiting from R&D activities and the building of economies of scale"* (page 14). However, this development cannot be observed anywhere in the real world. Nevertheless, retrofitting of existing plants and obliging new plants to be *"CCS-ready"* is addressed in addition. It will *"lead to*

---

[7]http://ec.europa.eu/energy/eepr/
[8]http://www.ner300.com/

higher costs and significantly reduced overall efficiencies than for newly built power plants with capture", and the *"number of plants that have been already designed 'CCS ready' is, however, very low"* (page 15).

The communication finds that carbon storage is facing *"strong public opposition"* (page 18), allegedly due to *"lack of public awareness"* – not taking into account that there are real health concerns for leakage problems and also public opposition against continued use of coal fired power plants, where CCS is not considered as an asset for climate protection but as a pretext for continued use of fossil fuels.

Building on the IEA's World Energy Outlook 2012, the communication expects increased deployment of coal fired power plants, particularly in China and other developing countries. As a consequence, the necessity of CCS is evoked using the same examples that were already used in 2009. *"Unless new plants in China and around the world can be equipped with CCS and existing plants retrofitted, a large proportion of the world's emissions between 2030 and 2050 are already 'locked-in'"* (page 19). And like all the years before, there is no indication given, why China and the other countries should consider using CCS, given the lack of an international climate regime – and given the lack of economic viability, which the communication correctly presents.

Based on the devastating analysis – no business case, much too expensive, lack of public acceptance – the Commission finds that *"CCS is now at crossroads"* (page 22) and asks for stakeholder views about new and additional policies to trigger CCS-deployment, including mandatory CCS certificates, feed-in tariffs and emissions performance standards. The outcome of this consultation is predictable. Most of those who have always advocated CCS will continue to do so in order to keep an option for continued use of fossil fuels. Like today, none of them really believes that China and other developing countries will seriously consider using CCS, nor will any of today's major economies. Most likely, increased deployment of renewable energy will be the only viable solution for significantly reducing greenhouse gas emissions, and certainly the only solution available in the foreseeable future.

11.7   EUROPEAN PARLIAMENT CALLING FOR A STABLE 2030-FRAMEWORK

Triggered by the Commission's Renewable Energy Strategy (European Commission 2012b), the European Parliament's ITRE-Committee decided to launch a report about *"Current challenges and opportunities for renewable energy on the European energy market"* (European Parliament 2012). The initiative came from German MEP Herbert Reul (Christian Democrats/EPP), who also provided the draft report. Reul, an explicit sceptic of renewables, publicly announced his report as calling for a harmonised European quota system to replace the existing different national support mechanisms (Welt 2013). When the draft report was published on 9 November 2012, the harmonised quota system was not the only imposition for supporters of the existing climate and energy framework and the 2020 targets.

The draft begins with statements most parliamentarians agree on, such as expecting increasing shares of renewables in the EU, which help reducing dependency from fossil fuels, or the necessity of supply security as well as economic and environmental viability. Then it turns to Reul's specific view on some issues. In his view *"approximately 170 different schemes for promoting RES give rise to considerable inefficiencies in cross-border electricity trading because it reinforces and indeed aggravates inequalities, thus working against completion of the internal energy market"* (European Parliament 2012, page 4), a statement which is a very special view on reality. There are not *"170 different schemes"* for renewables in Europe, nor are the different support schemes hindering the completion of the internal energy market. The opposite is true: due to market distortions and incomplete transposition of internal market legislation, support schemes are necessary to compensate for market failures. When discussing infrastructure requirements, the draft report rightly states that this has to be *"different from that currently in place"*, but then come up with highly exaggerated estimations about necessary *"reserves of conventional energy not previously available"* for balancing *"RES with fluctuating feed-in"*, assuming that *"development of reserve capacity entails substantial costs"*. Without solving the challenges for the network structure *"supply security is being severely affected by the increased feed-in from RES"* (page 5). Even concerns for Europe's landscape are raised when emphasizing *"that the further development of RES will entail permanent landscape change in Europe"* (page 6).

At the end of the draft, the author comes to his pre-announced intention, which is harmonising support schemes for renewables – an idea which has been rejected so many times by Parliament and Council. Based on his flawed assumption of 170 different *"types of promotion mechanisms"* the observation is made that *"support has led to healthy growth but that some of the promotion systems are very costly and that, in some cases, a considerable financial burden has been placed on consumers without their having had a choice in the matter"* (page 6). He notes that *"despite the subsidies, RES have managed to become competitive vis-à-vis conventional methods of energy production only in certain areas, e.g. where the geographical conditions favour them"* – ignoring that (including in the EU) the vast majority of subsidies still goes to fossil and nuclear energy, most of which would definitely not be competitive without.

Eventually, the point is made for harmonising support schemes by arguing that – although very much welcomed – *"good-practice guidelines are only a first step and that efforts need to be directed at winding down the national support systems"*. At least, he adds that *"they must not be retrospectively amended or cancelled because that would send out disastrous signals to investors"*, but he is *"convinced that only an EU-wide system for promoting RES will offer the most cost effective framework in which their full potential can be realised; sees decisive advantages in a technology-neutral European market for renewables, in which producers will have to cover a pre-determined quota of their energy output from RES, and in which one of the ways of reaching that quota will be through the trading of certificates on a market established for that purpose"*. All this does not comply with experience gained all over the EU, where technology neutral support mechanisms have gradually been replaced by technology specific systems, which have proven to be more effective and cost efficient. Nevertheless, the motion ends with calling *"on the Commission to bring forward, without delay, a proposal for a European support system in which a market for renewable-energy certificates will make for EU-wide competition among the various technologies"* (page 7).

The draft report provided evidence of how far some MEPs have moved away from the consensus of the climate and energy package of 2009. When it was finally voted in the in the ITRE Committee (European Parliament 2013), most of the biased and unsubstantiated statements had been replaced by more appropriate analysis and more realistic proposals for a post-2020 framework. With the support of NGOs and renewables associations, the shadow rapporteurs[9] of the ITRE-Committee had managed to draft and vote hundreds of controversial and eventually compromise amendments, which transformed the motion into a useful document for the debate about 2030-targets for renewables and an integrated climate and energy framework, still leaving some room for interpretation, but highlighting the need for a reliable post-2020 framework and targets to be set on the road to 2050.

After the Committee had voted and removed most of the controversial points, now clearly asking the European Commission to assess a binding 2030 renewable energy target, Herbert Reul underlined in a press communication that the Committee had called for a more European approach to renewables support with harmonised support criteria. Furthermore he commented on a motion from several MEPs to ask for a 45% renewables target for 2020, stating that *"more than doubling the share of renewables in such a short time would be completely utopian and in no way compatible with reliable and accountable policies"* (Reul 2013). His statement contrasted

---

[9]Reports in the EP are drafted by a *rapporteur* (and his or her assistants), who is supported by *shadow rapporteurs* (and their assistants) nominated by the other parliamentary groups. Rapporteur and shadow rapporteurs need to work closely together to prepare the voting, starting in the responsible committee and eventually going to the plenary of the European Parliament. A motion can only be adopted by majority vote in plenary.

with a press communication issued by the European Renewable Energy Council, welcoming that the Parliament *"calls for post-2020 renewable energy policies and targets to be set on the road towards 2050. This of course resonates with EREC's proposed 45% renewable energy target by 2030"*. In my capacity as EREC's President, I said *"This is a key message from the European Parliament to the Commission a few weeks before publication of its Green Paper on a 2030 policy framework for climate and energy"* (EREC 2013a).

Looking at the committee vote more closely reveals how much the original draft was changed, underlining once more that the European Parliament is strongly in favour of promoting and developing ambitious policies for renewable energy. The report acknowledges that *"the share in Europe's energy mix accounted for by renewable energy sources (RES) is growing in the short, medium and long term, and [. . .] RES make a significant contribution to guaranteeing a secure, independent, diversified and low emission energy supply for Europe"* (European Parliament 2013, page 4) and it highlights that *"investors require security and continuity for their projected investments beyond 2020"*. The Committee agrees *"with the Commission that RES, together with energy efficiency measures and flexible and smart infrastructure, are the 'no regrets' options identified by the Commission and that RES in the future will account for a growing share of energy provision in Europe, for electricity supply, for heating (which makes up nearly half of the total energy demand in the EU) and cooling and for the transport sector, and that they will reduce Europe's dependence on conventional energy"*. The Committee *"adds that targets and milestones should be set for the period to 2050 in order to ensure that RES have a credible future in the EU; recalls that all scenarios presented by the Commission in its Energy Roadmap 2050 assume a share of at least 30% RES in the EU's energy mix in 2030; suggests, therefore, that the EU should endeavour to achieve an even higher share"*. The report emphasizes *"that renewable energies not only contribute to addressing climate change and increase the energy independence of Europe but also offer significant additional environmental benefits through the reduction of air pollution, waste generation and water use, as well as of further risks inherent to other forms of power generation"* (page 5). The vote *"acknowledges the increasing competitiveness of renewable energy technologies and stresses that RES and clean-tech related industries are important growth drivers for Europe's competitiveness, representing an enormous job creation potential and making an important contribution to the development of new industries and export markets"*. The need for more renewables in the heating and cooling sector is highlighted, calling *"on the Commission and the Member States to pay more attention to the untapped potential of RES in the heating and cooling sector and to the interdependencies between and opportunities arising from increased renewable energy use on the one hand and the implementation of the Energy Efficiency and Buildings Directives on the other"* (page 6). Finding that a *"wide variety of different schemes for promoting RES currently coexist within the Union"* (page 8), the report *"welcomes guidance from the Commission on support scheme reform"* and on better use of cooperation mechanisms of the Renewables Directive. Concluding, the report summarizes the need for a *"European framework for the promotion of renewable energy"* and emphasizes the variety of support schemes pointing out that *"this support has led to healthy growth, in particular when support schemes are well designed, but that some of the promotion systems have been badly designed and have proved insufficiently flexible to adjust to the decreasing cost of some technologies and have, in some cases, caused overcompensation, thereby placing a financial burden on consumers without their having had a choice in the matter; is pleased to observe that, thanks to the subsidies, some RES have managed to become competitive vis-à-vis conventional methods of energy production in certain areas, e.g. where geographical conditions favour them, where access to capital is good, where the administrative burden is the lowest or by economies of scale"* (page 17). The need for stability and negative impact of retroactive changes to support frameworks is also included now. Looking beyond 2020, the report calls for moving the *"debate about greater convergence and a suitable European system of support for post-2020 forward"*. The report *"asks the Commission to assess, in the context of a post-2020 framework, whether an EU-wide mechanism for promoting RES would offer a more cost-effective framework in which their full potential could be realised, and how a progressive convergence could function"*. With regard to existing studies (e.g. REShaping

2010), it can be expected that the result of the assessment will be that strengthened national support systems are the most efficient way for promoting renewables.

The report, which MEP Reul had initiated in order to oppose and criticise existing renewable energy frameworks and to highlight problems of renewables integration – after debate and voting in committees (and presumably also the plenary[10]) of the European Parliament – has changed into a call for a reliable post-2020 framework with ambitious and realistic 2030-targets – a position which the European Commission could build on in the Green Paper on the future framework for Climate and Energy Policies (European Commission 2013a).

11.8   TOWARDS A 2030-FRAMEWORK FOR CLIMATE AND ENERGY POLICIES

On 27 March 2013, the European Commission published several documents analysing the imple-mentation of the 2020-framework and taking the debate about a post-2020 framework to a next stage. The *"Renewable energy progress report"* (European Commission 2013c) analyses Member States' progress towards the 2020 targets for renewable energy. In parallel, progress of Carbon Capture and Storage is assessed (European Commission 2013d) and a consultation is launched on how to deal with the limited progress so far (European Commission 2013e). The strategic document, the *"Green Paper – A 2030 framework for climate and energy policies"* (European Commission 2013a) addresses the scope and content of a post 2020 framework – including ques-tions about future targets, ambition levels, instruments, interaction between the different elements and about impacts on competitiveness.

Based on the latest available Eurostat data, which are based on Member States' progress reports submitted in 2011, the Commission finds that *"an impression is gained of a generally solid initial start at EU level but with slower than expected removal of key barriers to renewable energy growth, with additional efforts by particular Member States being necessary"*. The Commission finds that *"further efforts are needed in terms of administrative simplification and clarity of planning and permitting procedures and for infrastructure development and operation. And further efforts are needed regarding the treatment and inclusion of renewable energy production within the elec-tricity system. The general economic conditions in the EU today together with disruptive changes to support schemes for renewable energy (again, raising regulatory risk), add to the conclusion that further measures will be needed at Member State level in order to stay on the trajectory and for the targets to be achieved"* (European Commission 2013c, page 2). A considerable number of infringement procedures have been launched against Member States for non-transposition of the Renewables Directive in order to achieve *"the rigorous and complete implementation of Renew-able Energy Directive and commitments made in the National Renewable Energy Action Plans"* (page 13–14).

At the end of 2010, the share of renewable energy in the EU's energy mix was 12.7% (compared to the 2010 interim target of 10.7%). This is promising, although it has to be noted that the interim targets are more ambitious towards 2020 so that target achievement – without further measures – cannot yet be taken for granted, even less so as the 2010 figures do not include the impacts of the above mentioned retroactive and other changes to support mechanisms, which were enacted after 2010.

Looking at individual Member States, as in Annex I of the document, only Latvia (32.6% instead of 34%) and Malta (0.4% instead of 2%) were more than 1% below their 2010 interim tar-get. 13 countries (Austria, Bulgaria, Germany, Denmark, Estonia, Spain, Finland, Hungary, Italy, Lithuania, Romania, Sweden and Slovenia) had exceeded the interim target by more than 2%. Sweden had already exceeded the 2020-target of 49% by 0.1%, and Romania (23.6% compared to 24%) and Estonia (24.3% vs. 25%) were very close to the 2020-targets. Among the remaining

---

[10]The vote took place on 21 May 2013. In most major questions the ITRE-Version was endorsed. In addition, a clear demand for a binding 2030-renewables target was added by a (though small) majority of the Parliament (European Parliament 2013a).

countries, 10 had slightly exceeded the interim targets and two were below their respective trajectories (UK and the Netherlands). The figures and the analysis show that nearly all Member States need to increase their efforts to securely reach or exceed their 2020 targets.

11.8.1   *Green Paper: A 2030 framework for climate and energy policies*

The European Commissions seems to have understood the messages from the European Parliament and from stakeholders calling for a solid and reliable post-2020 framework. The Green Paper (European Commission 2013c) highlights three reasons for aiming at *"early agreement on the 2030 framework"* for climate and energy policies:

- *"First, long investment cycles mean that infrastructure funded in the near term will still be in place in 2030 and beyond and investors therefore need certainty and reduced regulatory risk.*
- *Second, clarifying the objectives for 2030 will support progress towards a competitive economy and a secure energy system by creating more demand for efficient and low carbon technologies and spurring research, development and innovation, which can create new opportunities for jobs and growth. This in turn reduces both directly and indirectly the economic cost.*
- *Third, while negotiations for a legally binding international agreement on climate mitigation have been difficult, an international agreement is still expected by the end of 2015. The EU will have to agree on a series of issues, including its own ambition level, in advance of this date in order to engage actively with other countries"* (page 2).

Whilst tackling the challenges ahead, the Commission underlines the need to analyse changes since the climate and energy package had been agreed on, in particular the economic and financial crisis, budgetary problems of some Member States, developments in the EU and in the global energy sector as well concerning questions about affordability of energy and impacts on competitiveness. Drawing lessons from the past, *"the framework should take into account the longer term perspective"* as laid out in the Low Carbon Roadmap 2050, the Energy Roadmap 2050 and the Transport White Paper, aiming at 40% greenhouse gas reduction in the EU by 2030 (and 80–95% by 2050) and at significantly higher shares of renewable energy, which was identified as a major *"no regrets option"*. As a result of the public consultation[11] the Commission plans to develop suggestions for a post-2020 framework – in the form of a *Communication,* a *White Paper* and/or legislative proposals, which will be discussed and assessed by the European Parliament in 2014 and most likely not be decided before 2015, when the newly elected Parliament and the next European Commission will have started their work. But it is definitely necessary to begin the process now in order to have major preparatory work done in 2014.

The Green Paper outlines the state of implementation regarding the different elements of the 2009 package.

Referring to overallocation of allowances in the EU-ETS, the paper finds that it *"has not succeeded in being a major driver towards long term low carbon investments. Despite the fact that the ETS emission cap decreases to around –21% by 2020 compared to 2005 and continues to decrease after 2020, in principle giving a legal guarantee that major low carbon investments will be needed, the current large surplus of allowances, caused in part by the economic crisis, prevents this from being reflected in the carbon price"*. The second element for achieving greenhouse gas reduction by 2020, the *"Effort Sharing Decision"* (see Chapter 3) to reach 10% greenhouse gas reduction outside the ETS, is also evaluated. Accordingly *"the EU is on track to achieve the 10% reduction target, but significant differences exist between Member States. Half of them still need to take additional measures"* (page 4).

For the progress in renewables development reference is made to the progress report, and to the need to reach an average annual growth rate of 6.3% to achieve the 2020-targets (up from 4.5% now, and only 1.9% in the period 2001–2010). Additional measures will therefore be needed to successfully progress towards 2020.

---

[11]The consultation will end on 2 July 2013 only. Therefore, the outcome cannot be analysed here.

For the energy savings target of 20% by 2020, which is not legally binding, progress is insufficient. With the Energy Efficiency Directive (Energy Efficiency Directive 2012) adopted in 2012, there is a comprehensive legal framework now, but with a *"lack of appropriate tools for monitoring progress and measuring impacts on the Member State level"* (European Commission 2013c, page 5) and without sufficient certainty to reach the 2020 target.

The Green Paper identifies four key issues for consultation, the first of which is about targets, relating *"to the types, nature and level of targets and how they interact. Should the targets be at EU, national or sectoral level and be legally binding?"* (page 7) The question is asked, whether the existing set of targets can be streamlined, e.g. by renouncing from a separate transport target, or by focussing on a greenhouse gas reduction target only. When it comes to a new renewable target, the green paper is ambiguous. *"Higher shares of renewable energy can deliver GHG reductions so long as these do not substitute other low-carbon energy sources"* (page 7), which is in line with other statements from the Commission, re-introducing nuclear as an alternative to renewable energy[12]. In addition, reference is made to interaction between renewable energy in ETS.

Setting the scene for what is to be analysed, the Green Paper presents several options. *"The Energy Roadmap for 2050 has shown that the share of renewables in the energy system must continue to increase after 2020. A 2030 target for renewables would have to be carefully considered as many renewables sources of energy in this time frame will no longer be in their infancy and will be competing increasingly with other low-carbon technologies. Consideration should also be given to whether an increased renewable share at EU level could be achieved without a specific target but by the ETS and regulatory measures to create the right market conditions"* (page 8).

The second issue is the *"coherence of policy instruments"* (page 9). *"The 2030 policy framework should, therefore, strike a balance between concrete implementing measures at EU level and Member States' flexibility to meet targets in ways which are most appropriate to national circumstances, while being consistent with the internal market"* (page 9).

The third element is *"fostering the competitiveness of the EU economy"* (page 10), which deals with development of energy prices as well as innovation in energy industry. There is a strong demand for fully implementing the internal energy market, but also a focus on the *"need to enable the future exploitation of indigenous oil and gas resources, both conventional and unconventional in an environmentally safe manner, as they could contribute to reducing the EU's energy prices and import dependence"* (page 11). And whereas the environmentally harmful search for unconventional sources is described as an – expensive – necessity, the fear is hedged of renewables increasing energy prices. Consequently, the narrative of cost savings through harmonised renewable energy support systems is kept alive.

As an overall assessment, the Green Paper raises questions, and is leaning towards problematic answers. Nevertheless, it opens a process, where the debate is opened about what kind of targets are needed for 2030 and about the framework needed. Stakeholders from the renewables sector and environmental NGOs will certainly take their chance to advocate a clear, ambitious and binding framework for increasing energy efficiency and the shares of renewable energy in Europe's energy mix.

11.8.2   EREC's Hat-trick 2030

The European renewable energy sector, represented by EREC, opened the new phase of the post-2020 debate by asking for a *"Hat-trick 2030. An integrated climate and energy framework"* (EREC 2013b), setting out a number of reasons for continued and ambitious renewables policies beyond 2020, in particular *"why an integrated renewables – greenhouse gas – energy efficiency 2030 policy approach with an ambitious and binding renewable energy targets yields more benefits for European citizens and industries than a one-legged policy based on a supposedly 'technology-neutral' GHG-only approach"* (page 5).

---

[12]For evidence see European Commission 2013, where *"market distortions"* are discussed, which might justify State Aid for nuclear energy.

Ten specific benefits of such an approach are laid out in detail. The most important benefit is *"Providing a clear signal for investors"* (page 8) and thus stabilizing the market alongside a long term priority for renewables, thereby creating *"a virtuous circle, where the expectation of sufficient infrastructure reduces the risks and thus the costs of renewable energy deployment"*. The integrated framework would help *"Growing the economy"* (page 9). Following the successful 2020-targets, which are expected to increase GDP by 0.25% until 2020, a stable 2030-framework *"could further increase to a minimum net GDP growth of 0.45%"*. The framework could reduce *"the costs of decarbonisation through both innovation and deployment"* (page 6), particularly if targeted support is developed for innovative technologies entering the markets and then reducing support when maturity increases. Research has shown that *"overall costs of buy-down will be much lower if policies target that specific market than if a generalised tool is used, such as carbon pricing"* (page 11). Thus a mix of *"demand pull (via markets created)"* and *"supply push (via R&D)"* helps to develop a broad technology mix and *"a timely scale-up of a wide set of renewable energy technologies"*.

EREC shows that an ambitious integrated framework with binding targets for renewables helps *"Reducing the costs of financing"* (page 12) by reducing *"political and regulatory uncertainty"* for investors, which is particularly important for financing of renewables with high upfront investment needs, such as wind and solar. Whereas costs of CO_2 will remain volatile, even with a functioning ETS, *"dedicated renewables policies offer a lower risk environment for investors, with regard to price risks, thereby lowering the costs of capital"* (page 13). Increasing the share of renewables by a targeted framework will directly support *"Reducing the need for support mechanisms"* (page 14) by narrowing the cost gap compared to other technologies, so that – provided markets are functioning properly – *"an increasing number of renewable energy technologies will be able to move away from existing support mechanisms"*.

Whereas other regions of the world, industrialised and emerging economies, are increasingly engaging in renewable energy development and deployment, without a clear framework, the EU risks losing the first mover advantage it has in clean technologies so far. Without a strong framework, supporting strong domestic markets, *"Enhancing EU technology leadership"* (page 15) will not be possible. With a clear commitment to further increasing the shares of renewables in the energy mix and dedicated policies in place, Europe could increase the renewables market volume despite decreasing shares in the global market. Technology leadership for system integration and system transformation would thus be highly beneficial for Europe's economy. Another benefit would be *"Reducing fossil fuel imports"* (page 18) and thus reducing trade deficit. EREC estimates that avoided fossil fuel imports would amount to € 388 billion, which is more than twice the present trade deficit of € 150 billion. A reliable framework would also help maintaining and *"Creating jobs"* (page 20), up from about 1.1 million to 2.7 million in 2020 and 3.6–4.4 million in 2030.

The *"hat-trick"* will help *"Protecting the environment"* (page 22) by reducing carbon dioxide and other emissions, such as methane, sulphur dioxide, ozone, etc. and in particular emissions from coal-fired power plants and the resulting health costs of up to € 43 billion a year in the EU. And it would lower water consumption and thus save scarce resources. Eventually, developing a broad range of renewables technologies will support EU Member States in choosing and focussing on those technologies and resources, which are best available and manageable in the respective regions.

With the *"Hat-trick 2030"*, the European renewable energy sector has presented a comprehensive paper summarizing the most important arguments for the debate about a post-2020 framework for the EU, following up on the climate and energy package of 2009 which established the globally recognised 20-20-20-targets for 2020. It is obvious that a follow-up is needed to avoid expiring of the policy framework and reduced momentum with reduced ambition when it comes to moving towards a fully sustainable energy future for Europe.

Outlook – towards 100% renewable energy

Rainer Hinrichs-Rahlwes

1 PAVING THE ROAD TOWARDS A TRULY SUSTAINABLE ENERGY SYSTEM IN EUROPE – DEVELOPING A 2030 FRAMEWORK

A set of 2020-targets and policies are in place. A comprehensive framework for renewable energy and greenhouse gas reduction was set up and enacted. And – with some relevant exceptions – it is being implemented. It has become a major pillar of Europe's policy priorities, with all its different elements, with its strengths and weaknesses, from the Renewables Directive with binding targets for each Member State, via the improved but still not effectively working Emissions Trading Directive to the Effort Sharing Decision and the non-binding efficiency target to the CCS-Directive, which is trying to develop a technology which will most likely never become economically viable and, eventually, the greenhouse gas reduction target of 20%, which definitely should but probably will not be raised to 30%, because there will not be an international agreement with similar ambition. The framework as it was agreed in 2008/2009, together with the single market legislation, is the starting point for further development beyond 2020. It provides guiding principles for a process of setting up a next milestone for Europe's climate and energy policies – including policies, a regulatory framework and ambitious targets.

The overall positive assessment of the framework is put in question by some Member States who are shifting from ambitious and full implementation to revising their national targets and policies in a way that achieving the 2020-targets is seriously at risk. This is a major stumbling stone to be taken into account when it comes to setting up the framework for the next decade. Some Member States remain strongly dependent from their incumbent utilities. They still have markets and energy systems in place, which were designed for conventional and/or nuclear energy and are still being operated accordingly. As a result, subsidies for conventional energy continue to be much higher than those for renewables, but nevertheless public opinion is lead to discussing increasing energy prices allegedly due to renewables support. If existing and written off conventional power plants are compared to new renewables, conventional energy seems to be cheaper. If societal costs of nuclear are borne by taxpayers instead being included in the electricity prices, nuclear power appears to be cheaper than some renewables. And with a dysfunctional ETS, coal fired power plants seem to be cheaper than they actually are. Strong incumbents have their means to spread their views of energy security and affordability via media, sometimes heavily depending on their advertising.

Support schemes and supportive framework conditions with ambitious targets are necessary to further accelerate development and deployment of renewable energies. It is indispensable that decisions to phase out fossil fuel subsidies are implemented in order to create a level playing field with increasingly internalised external cost of fossil fuel production and consumption. In this context, support for Carbon Capture and Storage is highly questionable, because the main purpose of CCS, as it is discussed today, seems to be perpetuating the use of coal, oil and gas under the headline of low-carbon energies[1]. It will be of particular importance to limit public spending

---

[1] It is not necessary to discuss in detail here whether or not – in the context of an energy system with very high shares of renewable energy – it might be useful or necessary to capture carbon emissions from biomass in order to achieve additional greenhouse gas reductions, which are necessary to limit global warming. In the present discussion, this is not a relevant option. And safe storage of carbon dioxide would still be an issue to be solved, before "biomass-CCS" could be accepted under sustainability aspects. Alternatively, material use of the captured gases could be considered.

for this immature and potentially problematic technology, which is predominantly promoted by those who plan to continue using high quantities of domestic coal for their energy supply. As long as no consensus is found for revitalising the ETS and transforming it to become an effective instrument for carbon pricing, there will not be any economic rationale for CCS development, which is significantly more expensive than the price for carbon allowances in a dysfunctional carbon market. And although the European Commission, in its recent consultation document on the future of CCS (European Commission 2013d) reiterated the argument of China and other developing countries needing CCS, because they will continue to use coal, the perspectives of economic viability for CCS do not become brighter than they were a few years ago. Without a global carbon market with meaningful carbon prices, CCS is not a real-life option. And without the European Union leading by example, a functioning global carbon market is unlikely to develop.

Another deviation on the way towards a sustainable energy future is pursued by some stakeholders and some governments, which have re-boosted their activities to promote nuclear as a *"low carbon energy source"*, a classification which even made it into European Commission Documents (e.g. European Commission 2013, European Commission 2013a). This is not new, but it is more irresponsible after the nuclear disaster in Fukushima than it was before, due to unsolved security problems and the persisting absence of sustainable solutions for nuclear waste disposal or treatment. Unfortunately, the nuclear option has friends in some European governments and in the European Commission. Although nuclear power has been and still is heavily subsidised in the last seven decades and although nowhere in the world nuclear power plants have to pay insures fees comparable to those of other energy producing facilities, Finland, Poland, the Czech Republic and the United Kingdom are trying to (re-)introduce new nuclear into their energy mix. In Finland, a new reactor (Olkiluoto 3) is being built. Instead of starting operation in 2009, as originally planned, this is now foreseen for 2015. And cost estimates have increased from originally € 3 billion to nearly 9 billion now, with no upper limit foreseeable yet. Poland and the Czech Republic are trying to design frameworks for new reactors to become economically viable (with massive state guarantees, or even with direct state aid).

In the UK, a law is under consideration in Parliament to introduce – via so called contracts for difference – minimum prices for low carbon energies including nuclear. The UK (only 3.3% RE in 2010, European Commission 2013c) is far below the trajectory towards the 2020 renewables target and strongly opposing new renewables targets for 2030. But they are promoting the introduction of guaranteed prices of more than 10 €-cent/kWh for nuclear, for a period up to 40 years. And still it seems they do not find an investor who is willing to take the risk. This could be a major reason, why the European Commission's DG Competition is analysing whether there are *"market failures"* to the detriment of nuclear energy so that State Aid would be acceptable and lawful in order to support *"advancing towards a low-carbon energy system"*. It remains to be seen, if they really try crafting and imposing a framework where nuclear receives support against market distortion. This would counteract all the arguments about only supporting *"infant technologies"*. But it is what the UK and some others are pushing for in present debates. Looking at actual developments, it seems that – without massive new subsidies, which are probably unlawful with respect to European State Aid regulation – no new nuclear power plant will be built in Europe, where financing (and probably also public opinion) really is an issue.

With nuclear and CCS being too costly and too risky, developing and deploying renewables (together with increased energy efficiency and energy saving) is the only realistic option for greenhouse gas reduction that is actually available. Political reality in the EU, however, is partly different from this obvious truth.

It has been shown in Chapter 7 that both in the Low Carbon Roadmap 2050 and in the Energy Roadmap 2050, irrespective of the detailed assumptions and settings, all scenarios complying with the greenhouse gas reduction targets in Europe of 80–95% by 2050 have very high shares of renewables in the 2050 energy mix, even those, where nuclear and CCS are (by definition) contributing significantly to the energy supply. Based on these results, a policy process has to be further developed to agree on new milestones. Certainly, a consensus about a target of 80 or 100% renewables by 2050 is not a realistic policy option at present (although it would be useful and technically and economically feasible), but a framework can and must be developed, which is in line with the greenhouse gas reduction targets. And this should be easier when it comes to defining shorter term milestones, which do not require further reaching consensus among the Member States. Keeping and further strengthening the momentum created by the 2020 framework will be an important asset in this context.

The endeavour to move forward towards a new and ambitious framework for the post-2020 decade will not be easy, but it can be done. The willingness to discuss 2030 milestones needs to be transformed into a debate and then a consensus about an integrated framework of targets and related enabling measures. The targets need to be clear and ambitious – and they must not distract major policy decisions from maintaining and further developing a stable and reliable framework for renewables. Although there will probably not be a consensus not to support CCS and not to use nuclear, it could be possible (as it was in 2009 for 2020) to reach a consensus about facilitating the no-regrets options renewables, efficiency and infrastructure development. This consensus will have to be based on thorough analysis of costs and benefits of such a new framework.

An important precondition will probably be an in-depth assessment of costs and benefits of different options, which provides credible and acceptable data for most of the EU Member States, and which results in an informed discussion about the future of support systems for renewables in the EU. Implications of increasing the shares of renewables on energy market designs and vice-versa must be part of the process. As outlined in Chapter 11, results are not likely to be agreed on before the newly elected European Parliament and the next European Commission will have taken office in 2014. It may even take until 2015, before agreement can be reached. Until then, the interlinked debates about 2030 milestones and/or targets and about the future of renewables support systems will have to result in substantial and consensual findings.

2 AN INTEGRATED 2030-FRAMEWORK FOR RENEWABLES, EFFICIENCY AND GREENHOUSE GAS REDUCTION

Support schemes have never been an end in themselves, but they are tools for achieving defined targets or ambition levels. This is why early agreement on a stable and predictable policy frame-work for the decade after 2020 is so important. Such an agreement on a new set of targets and framework conditions will provide guidance for investors by outlining minimum market size and minimum growth rates – irrespective of the possibility for Member States to pursue more ambitious goals by additional targeted action. I have described in Chapter 11, how the European Commission has set the scene by opening a public consultation on the future framework. Given the controversies between those who are promoting a mix of *"low carbon energies"* and those who are promoting sustainable renewable energy, it is important to avoid potential traps, e.g. by fixing a minimum consensus, which consists of greenhouse gas reduction targets only. With such a consensus reached, discussions about new binding (and ambitious!) targets for renewables and efficiency would actually be void. These specific targets would then be treated as inferior to the overall target of greenhouse gas reduction. And those who would like to step away from renewables and instead burn money on coal and CCS would celebrate, because they could claim to do so in compliance with an overarching EU-target.

Legally, an agreement on 2030 greenhouse gas reduction targets could be voted with a qualified majority in the Council, whereas some argue that a new renewables target would need unanimity. This cannot be a striking argument, however. We have seen in 2009 (and the years after) that unanimity was politically indispensable to reach an agreement on the climate an energy package, although – legally – this might not have been necessary. On the other hand, even a unanimously agreed framework can be questioned afterwards. It is therefore of paramount importance to craft a 2030 framework which is designed to be agreed on by every Member State and which includes a certain level of commitment to be implemented in reality. An agreement on new greenhouse gas reduction targets alone, without underpinning them with efficiency and renewables targets, would probably not be implemented, because it would not include concrete enough measures and subtargets that can be concretely monitored. Given the present discussions, a greenhouse-gas-only target would be a wildcard for those who want to get rid of renewables obligations – either to support new nuclear instead and/or to get rid of any kind of binding greenhouse gas reduction obligation.

A new set of ambitious renewables and efficiency targets would not immediately stop the debate about ambition levels and nuclear renaissance, but advancing within such an agreed framework of binding targets for renewables and efficiency would deliver a significant greenhouse gas reduction simply through implementation of the targets, irrespective of whether the greenhouse gas target reduction target itself is binding or not. And the specific targets would not only create a stable and reliable framework for investors, but they would also decrease the costs of greenhouse gas mitigation and technology development by reducing technology risks. The challenge for the coming years will be to promote and successfully achieve an ambitious and binding renewables target of 45% for 2030, combined with a binding efficiency target of the same ambition level. These are objectives which can be implemented and achieved by domestic EU action. Based on these domestically achievable binding targets, an agreement about the level of greenhouse gas reduction in the EU by 2030 can be developed, starting from the reductions provided by efficiency gains and by renewables and adding those which have to be achieved by other instruments in other sectors.

3   THE FUTURE OF SUPPORT SCHEMES

The ambition level of a 2030 renewables target is an important indicator for the future design of support schemes for renewables. Ambitious objectives underpinned by binding targets would require more targeted support than a low ambition level close to business as usual projections. At least, consensus achieved about the ambition level for 2030 and about effort sharing would provide valuable criteria for support scheme design. To achieve further significant increase of renewables deployment, minimum elements of support scheme design need to be secured. To deliver on similar growth rates as were observed so far and which will be necessary to achieve the 2020-targets, a stable and reliable framework will be needed, with technology specific support and smart balancing of market options and secure income for – in particular – independent producers of renewable energy.

It has been shown that well designed renewables support schemes have triggered unprecedented growth with hundreds of thousands of new jobs and significant economic growth. This has created and stabilized – in an increasing number of countries in Europe and world-wide – independent producers of energy, competing with the incumbents for market access and market shares. Where support schemes are providing stability for a number or years and where they are technology specific and regularly adapted to cost development and market changes, and particularly, where they incentivise cost reduction, they have created domestic jobs and manufacturing industries of considerable size. But some support schemes are badly designed and therefore not providing the necessary stability for investors – for example due to frequent or even retrospective changes, too high or too low remuneration, or a lack of differentiation, or due to administrative barriers which delay or prevent renewables projects. These bad examples are used to publicly discredit renewables (instead of blaming the bad systems) and accusing them for driving up costs of energy. It seems that a consensus is developing, underlined by upcoming guidance documents on support scheme reform announced by the European Commission for mid-2013, and also by the IEA (IEA 2013), that predictability of policies and frameworks is necessary for investors' confidence and thus for reducing costs of finance due to reduced risks.

With more and more renewables technologies becoming cost competitive with conventional energies in an increasing number of regions, even in distorted markets, it is obvious that support schemes have successfully spurred market penetration. In Denmark, Germany, Spain and increasingly in other countries, wind, solar and other renewables have become important if not dominant energy sources. Support schemes – which should rather be called remuneration schemes, because they define and guarantee a fair remuneration for new technologies in imperfect markets – must now be further developed and gradually adapted to increasing shares of renewables and increased cost competitiveness of more and more technologies in an increasing number of markets. Maintaining regular revisions in order to avoid overcompensation and facilitate cost decreases will continue to be key aspects for fine-tuning remuneration systems. As long as markets are disturbed, however, dominated by strong incumbents and unable to deliver price signals for a system based on variable renewables, priority (or at least guaranteed) grid access and dispatch will have to be preserved. Incentives should be developed to guide more and more technologies closer to market risks (and develop markets so that they can accommodate high shares of renewables).

Still, there is a broad variety of market designs and of support schemes so that it is impossible to impose a single mechanism all over the EU. However, it is interesting to observe that – despite all practical and theoretical differences – major features of support schemes are converging. Most countries with quota and certificate systems have introduced technology specific elements. They use "banding" for differentiation, i.e. different numbers of certificates are awarded to different technologies, according to different sites or to different technology costs. Some have even introduced feed-in support for less mature and/or small scale technologies. On the other hand, temporary opt-out is offered as a choice in most feed-in systems. Fixed price feed-in systems (feed-in tariffs) were supplemented with premium options, where producers can choose between the fixed tariffs or the market price with a premium on top. In premium systems (FIP or feed-in premium) lower cost fixed price options were introduced as a choice for investors with lower risk tolerance. All these mechanisms can be and in some countries are combined with prior auctions to determine the eligible projects in a bidding process. On the other hand, pure bidding systems are no longer in place, because they were assessed to be less appropriate for targeted and cost effective renewables support.

The process of learning from good practise is discussed as *"convergence"* of support schemes. Evidence shows that there is no need – and not even a real case – for top down *"harmonisation"* of support schemes. The opposite is true.

Where policies are designed to facilitate smooth and stable growth of renewables, minimum standards are developing as nuclei of a future European support framework. These elements are:

- stability over a certain period of time (10, 15 or more years),
- regular revision (every two or three years), but not too frequently and not including retroactive or retrospective changes,
- incentives for cost reduction (e.g. through degressive remuneration for new installations),
- technology and resource specific remuneration,
- different remuneration according to size and site of the installation,
- availability of finance, and costs of finance, which includes country risks, which are another element to be considered and included in support scheme design.

The Commission Green Paper about a 2030 framework for climate and energy policies includes a broad range of questions about the future of support schemes for renewables. They need to be analysed, discussed and answered in a constructive and non-disruptive way. And although convergence between various systems can be observed as a matter of fact, it is still necessary to smoothly develop joint elements and joint features in a way that respects the differences between the Member States and their energy sectors and which takes into account different intentions and ambitions behind certain support scheme designs. Precondition for future good practise certainly is the complete implementation of existing legislation as in the Renewables Directive and in the Internal Energy Market packages. Over time it is likely that more features of support for renewables in the different Member States will converge. This will, however, depend on whether and to which degree an agreement can be found about an integrated 2030 framework. And it has to be a process on a guided but voluntary basis, where the Commission's primary role must be to insist on the implementation of existing legislation and to support Member States, when they seek assistance for targeted closer cooperation. Secondly, guidance for support scheme reform with regard to developing technologies and markets should be provided. The respective documents announced by the Commission can be useful tools to this end, just as projects conducted by scientists and stakeholders can provide valuable expertise for policy development in this regard.

4   POLICY DECISIONS TO BE TAKEN

Timely development of supporting framework conditions for renewables, particularly the Electricity Directive 2001, the Biofuels Directive 2003, and eventually the climate and energy package of 2009 with the Renewables Directive as the flagship piece, was a key success factor for Europe's frontrunner role in greenhouse gas reduction and renewables development. This is a good starting point to move forward to developing and implementing a truly sustainable energy system. Scenarios – including the Commission's Energy Roadmap 2050 – underline that ambitious decarbonisation goals can be met at reasonable costs – and that renewables will be the main energy sources in all decarbonisation scenarios. However, decarbonisation of the energy sector will not happen automatically. Policy decisions have to be taken to facilitate the necessary developments. These decisions include, but are not limited to a 2030 framework with ambitious and binding renewables and efficiency targets.

Completing the internal market, in particular limiting the dominance of incumbents and establishing really independent – i.e. fully unbundled – system operators, continues to be of key importance in the quest for a level playing field with fully internalised external costs of energy. Complete phase out of direct and indirect subsidies for fossil fuels and for nuclear is another important requirement for effective removal of competitive advantages for unsustainable energy production. Fair energy and/or emissions taxation would be another asset. Developing larger balancing areas in electricity markets would help wind and solar to tap their full potentials, including participation in flexibility markets. Maintaining priority for renewables as long as market distortions are not removed is another central building block for a fully renewables based energy future.

5   FROM INTEGRATION OF RENEWABLES TO SYSTEM TRANSFORMATION
FOR RENEWABLES

With renewable energies becoming major players markets and grid systems need to follow this development. Small shares of renewables can easily be integrated in nearly every energy system due to a limited impact on load and demand flows and thus on grid stability and market functioning. With higher penetration, however, particularly with effective priority access in place, grids and markets face some challenges to be overcome on the way towards 100% renewable energy.

Conventional energy systems mainly rely on "baseload" powerplants – technically designed and economically calculated to deliver energy around the clock. In most countries, baseload is provided by coal or nuclear power plants. To ramp them up and down takes at least hours or more, and it can endanger the integrity and safety of the installation. Economically they are calculated assuming high load factors, electricity prices deriving from these high load factors plus the fuel costs. For baseload plants the lack of flexibility is not a problem, but it is part of the design. Regularly needed additional power is sourced from mid-merit power plants, often gas fired, or also coal. Eventually, peak demand is sourced from flexible "peak load" power plants, frequently natural gas fired. These installations have to be technically flexible so that they can be ramped up at short notice, when needed, and ramped down quickly, when the demand is reduced again. This capability of peak load plants is important for load following – for sudden increase e.g. during half-time breaks of televised football matches or for sudden decrease, when a big load suddenly stops operating (e.g. unforeseen shut down of machines). Peak load is usually traded at specific stock exchange markets, where prices follow the merit order – the cheapest power plant sells first and the most expensive one only after the cheaper ones have reached their availability limits[2]. Particularly gas fired power plants are calculated with lower capacity factors stemming from their role in the merit order. Technically, gas turbines are flexible so that they are ideal for providing peak load whenever needed.

With increasing capacities of wind and solar and resulting high shares of variable power production, the paradigm of baseload and peakload becomes a burden to system stability. With strong wind blowing and/or bright sun shining, conventional power plants have to be ramped down or disconnected from the grid to avoid overload. Albeit technically feasible, it reduces their full load hours. As a result, the economic viability is reduced or threatened. Refinancing becomes more costly, because the amortisation takes longer (and/or the costs per MWh increase), which reduces competitiveness with regard to other technologies. For coal and nuclear power plants, this is – primarily – a technical problem, because the output is not variable enough to comply with a flexible power system. For traditional peak load plants like natural gas driven turbines, the technical challenge can be met, but the economic challenges become more demanding. With higher shares of wind and solar, which have no fuel costs and therefore (nearly) no marginal costs (because wind and sun are resources free of charge, in contrast to gas, coal, oil and also biomass), electricity prices go down. In particular, at the peak load stock exchange, high shares of wind and solar provide more low cost energy supply. As a result, expensive gas fired power plants are no longer needed for system stability in a given moment and/or too expensive to remain in the merit order, so that their capacity factors are further reduced, lowering their economic viability. Therefore flexible gas fired power plants might not be available, when their flexible outeput is needed for system stability.

These challenges – limited flexibility of baseload plants and reduced profitability of gas fired power plants – result in the paradox effect that flexible power plants, which would be needed for system stability reasons, have to be disconnected from the grid, because the inflexible ones cannot be ramped down or disconnected for technical reasons. As a result, flexible power plants are no longer economically viable in the existing market design. This increases the costs of system integration of wind and solar. They have to be curtailed for system stability reasons, which either reduces their economic viability or it increases electricity prices, when curtailment is financially compensated. Alternatively, more power lines have to be built to deal with the increasing amounts of wind and solar, while inflexible baseload is kept running, needing grid capacity even at times, when their power is not needed for security of supply.

The answer to this challenge is preparing for the next step forward – facilitating the paradigm shift towards a flexibility driven energy system. This means moving from a baseload driven inflexible system towards a smart system with wind and solar becoming the *"new baseload"* or – more precisely – the new foundation of the energy system. Instead of merely trying to integrate solar and wind into an energy system which was developed for another paradigm (baseload and incremental peak load), the way forward is replacing baseload power plants by flexible power plants and combining this with other flexibility options such as demand side management, load shifting or innovative approaches, which have in common that they provide flexibility in a renewables based energy system (Feature *"System Transformation"* in GSR 2013).

A flexibility driven system based on variable renewables is the way forward for a future oriented energy system in many countries and regions. For reasons of climate protection, energy security, innovation and scarcity of fossil resources we need to build our future energy demand on renewable energy. No alternative is in sight to establishing a flexibility driven system. Variable wind and solar together with (partly) dispatchable hydropower, geothermal and biomass, is balanced by flexible power plants (natural gas as long as there is not enough biogas and/or other flexibility technology in the system), smart grids, large scale interconnection, centralised and decentralised storage as well as demand response and deeper integration of the three sectors. Excess electricity can be used for heating, for hydrogen production or can be exported or stored. Electric vehicles

---

[2]The merit order and the merit order effect are analyzed e.g. in IEA 2011 and in BEE 2011.

can serve as temporary storage capacities while they are parked and grid connected. The gas grid can become a huge storage system, if biogas and hydrogen are injected and used for heating, transport and electricity production, when needed.

Properly designed energy markets can deliver the services which are needed for a flexible system: electricity, heating, cooling, transport fuels and flexibility options – through flexible power plants and through shifting of load and demand. A smart system needs innovative technology. And a smart system integrates decentralized energy production and consumption with large scale balancing areas for mitigating supply and demand variability. In such a system the full potentials of renewable energy can be tapped – millions of jobs will be created and secured, and energy costs will remain reliable, because they become increasingly independent from fossil fuel imports and related price volatilities. Scenarios have shown that 100% renewable energy is achievable and that this is a solid basis for greenhouse gas reduction and climate protection.

Several steps have been taken already. This book has analysed the progress made so far and the solutions yet to be found and enacted in the EU. Recent debates as well as pending decisions are described. The process towards a 2030 framework for climate and energy policy needs to be driven forward now. If the European Union agrees on a new and effective policy framework with ambitious and binding 2030-targets for renewables, efficiency and greenhouse gas reduction, a major step forward will have been done. If in addition subsidies for fossil and nuclear energy are finally phased out and a level playing in the energy market is becoming reality, this would mark the point of no return towards a renewables based sustainable energy supply in Europe – with first mover advantage for Europe's economy and serving as a strong example for other parts of the world to pursue their own ambitions towards a clean and sustainable energy future.

References

Agora 2013: Agora Energiewende, 12 Insights on Germany's Energiewende, A Discussion Paper Exploring Key Challenges for the Power Sector, Berlin Februar 2013, http://www.agora-energiewende.de/fileadmin/downloads/publikationen/Agora_12_Insights_on_Germanys_Energiewende_web.pdf

Barents 1990: René Barents, The Community and the Unity of the Common Market, GYIL 33 (1990), pp. 9–36.

BBH 2011: Dr. Dörte Fouqet and Amber Sharick, Becker Büttner Held (BBH), prepared for the project Renewable Energy Policy Action Paving the Way towards 2020, Meeting the Renewable Energy Policy Mandate in 2020, Policy Recommendations & Best Practices from the EU Member State National Action Plans, Brussels 2011, http://www.repap2020.eu/fileadmin/user_upload/Events-docs/Brochures/REPAP2020_Policy_Recommendation_Paper.pdf

BEE 2011: Dietmar Schütz, Björn Klusmann (Herausgeber), Die Zukunft des Strommarktes, Anregungen für den Weg zu 100 Prozent Erneuerbare Energien, Bochum 2011, http://www.bee-ev.de/_downloads/energieversorgung/1106_BEE-Sammelband-Strommarkt-SI-9.pdf

BEE 2012: Uwe Leprich et al., Kompassstudie Marktdesign, Leitideen für ein Design eines Strommarktsystems mit hohem Anteil fluktuierender Erneuerbarer Energien, Bochum, Dezember 2012, http://www.bee-ev.de/_downloads/publikationen/studien/2012/1212_BEE-GPE-IZES-Kompassstudie-Marktdesign.pdf

Biofuels Directive 2003: Official Journal of the European Union, Directive 2003/30/EC of the European Parliament and of the Council of 8 May 2003 on the promotion of the use of biofuels or other renewable fuels for transport, 17.05.2003, L123/42, http://eur-lex.europa.eu/LexUriServ/LexUriServ.do?uri=OJ:L:2003:123:0042:0042:EN:PDF

CEER 2007a: ERGEG, 3rd Legislative Package Input, Paper 1: Unbundling, Ref: C07-SER-13-06-1-PD, 5 June 2007, http://www.energy-regulators.eu/portal/page/portal/EER_HOME/EER_PUBLICATIONS/CEER_PAPERS/Cross-Sectoral/2007/C07-SER-13-06-1-PD_3rdLegPackage_Unbundling_final_0.pdf

CEER 2007b: ERGEG, 3rd Legislative Package Input, Paper 2: Legal and regulatory framework for a European system of energy regulation, Ref: C07-SER-13-06-02-PD, 5 June 2007, http://www.energy-regulators.eu/portal/page/portal/EER_HOME/EER_PUBLICATIONS/CEER_PAPERS/Cross-Sectoral/2007/C07-SER-13-06-2-PD_3rdLegPackage_EnergyReg_final.pdf

CEER 2007c: ERGEG, 3rd Legislative Package Input: Paper 3: Network Regulation – Overall Framework, Ref: C07-SER-13-06-3-PD, 5 June 2007, http://www.energy-regulators.eu/portal/page/portal/EER_HOME/EER_PUBLICATIONS/CEER_PAPERS/Cross-Sectoral/2007/C07-SER-13-06-3-PD_3rdLegPackage_Network_Regulation_fina.pdf

CEER 2007d: ERGEG, 3rd Legislative Package Input, Paper 4: ETSOplus/GIEplus, Ref: C07-SER-13-06-4-PD, 5 June 2007, http://www.energy-regulators.eu/portal/page/portal/EER_HOME/EER_PUBLICATIONS/CEER_PAPERS/Cross-Sectoral/2007/C07-SER-13-06-4-PD_3rdLegPackage_ETSO-GIE_final.pdf

CEER 2007e: ERGEG, 3rd Legislative Packe Input, Paper 5: Powers and Independence of National Regulators, Ref: C07-SER-13-06-5-PD, 5 June 2007, http://www.energy-regulators.eu/portal/page/portal/EER_HOME/EER_PUBLICATIONS/CEER_PAPERS/Cross-Sectoral/2007/C07-SER-13-06-5-PD_3rdLegPackage_PowersNRA_final.pdf

CEER 2007f: ERGEG, 3rd Legislative Package Input, Paper 6: Transparency requirements for Electricity and Gas – a coordinated approach, Ref: C07-SER-13-06-6-PD, 5 June 2007, http://www.energy-regulators.eu/portal/page/portal/EER_HOME/ EER_PUBLICATIONS/CEER_PAPERS/Cross-Sectoral/2007/C07-SER-13-06-6-PD_3rdLegPackage_Transparency_final-2.pdf

CCS-Directive 2009: Official Jounal of the European Union, Directive 2009/31/EC of the European Parliament and of the Council of 23 April 2009 on the geological storage of carbon dioxide and amending Council Directive 85/337/EEC, European Parliament and Council Directives 2000/60/EC, 2001/80/EC, 2004/35/EC, 2006/12/EC, 2008/1/EC and Regulation (EC) No

1013/2006, 05.06.2009, L140/114, http://eur-lex.europa.eu/LexUriServ/LexUriServ.do?uri=OJ:L: 2009:140: 0114:0135:EN:PDF

Community Guidelines 2008: Official Journal of the European Union, Community guidelines on State aid for environmental protection, 1.4.2008, C 82/1, http://eur-lex.europa.eu/LexUriServ/ LexUriServ.do?uri=OJ:C:2008:082:0001:0033:EN:PDF

Council 1986: Official Journal of the European Communities, Council Resolution of 16 September 1986 concerning new Community energy policy objectives for 1995 and convergence of the policies of the Member States, 25.9.1986, C241/1, http://eur-lex.europa.eu/LexUriServ/LexUriServ. do?uri=OJ:C:1986:241: 0001:0003:EN:PDF

Council 1993: Official Journal of the European Communities, Council Decision of 13 September 1993 concerning the promotion of renewable energy sources in the Community (Altener Programme), Brussels, 18.9.1993, L235/41, http://eur-lex.europa.eu/LexUriServ/LexUriServ.do?uri=OJ:L:1993: 235: FULL:EN:PDF

Council 2007a: Brussels European Council, 8/9 March 2007, Presidency Conclusions, http://www. consilium.europa.eu/uedocs/cms_Data/docs/pressdata/en/ec/93135.pdf

Council 2007b: Press Release 12.03.2007, The Spring European Council: integrated climate protection and energy policy, progress on the Lisbon Strategy, http://www.eu2007.de/en/News/Press_Releases/ March/0312AAER.html

Council 2011a: European Council 4 February 2011, Conclusions, http://register.consilium.europa.eu/ pdf/en/11/st00/st00002-re01.en11.pdf

Council 2011b: Council of the European Union, Council conclusions on strengthening the external dimension of the EU energy policy, 3127th TRANSPORT, TELECOMMUNICATIONS and ENERGY Council meeting (Energy items), Brussels, 24 November 2011, http://www.consilium. europa.eu/uedocs/cms_data/docs/pressdata/en/trans/126327.pdf

Council 2012a: Council of the European Union, European Council 1/2 March 2012, Conclusions, http://www.consilium.europa.eu/uedocs/cms_data/docs/pressdata/en/ec/128520.pdf

Council 2012b: Council of the European Union, Council Conclusions on Renewable Energy, 3204th Transport, Telecommunications and Energy Council meeting, Brussels, 3 December 2012, http://www.consilium.europa.eu/uedocs/cms_data/docs/pressdata/en/trans/133950.pdf

Dreher 1999: Meinrad Dreher, Wettbewerb oder Vereinheitlichung der Rechtsordnungen in Europa?, JZ 1999, pp. 105–112, http://www.jstor.org/discover/10.2307/20824783?uid=3737864&uid=2129& uid=2&uid=70&uid=4&sid=21101970774063

ECF 2010: European Climate Foundation, Roadmap 2050 – A Practical Guide to a Prosperous, Low-carbon Europe, 3 Volumes, April 2010, http://www.roadmap2050.eu/

ECJ 1978: European Court of Justice, Ramel C-80 and 81/77, Coll. 1978, 927, http://curia.europa.eu/jcms/jcms/j_6/

ECJ 1979: European Court of Justice, Rewe Zentral AG ./. Bundesmonopolverwaltung für Branntwein (Federal Monopoly Administration for Spirits), C-120/78, Coll. 1979, p. 649 Para. 8 and 14-Cassis de Dijon, http://curia.europa.eu/jcms/jcms/j_6/

ECJ 1984: European Court of Justice, Denkavit Nederland, C-15/83, Coll. 1984, 2171, http://curia.europa.eu/jcms/jcms/j_6/

ECJ 1984a: European Court of Justice, Rewe, Cases 37/83, Coll. 1984, 1229, http://curia.europa.eu/jcms/jcms/j_6/

ECJ 1984b: European Court of Justice, Luigi and Carbone, Joined Cases 286/82 and 26/83, Coll. 1984, 377, http://curia.europa.eu/jcms/jcms/j_6/

ECJ 1985: European Court of Justice, Association de défense des brûleurs d'huilesusagées (ADBHU), C-240/83, Coll. 1985, 531; http://curia.europa.eu/jcms/jcms/j_6/

ECJ 1988: Dänische Pfandflaschen (Danish returnable bottles), C-302/86, Coll. 1988, 4607 Para. 6, 9, http://curia.europa.eu/jcms/jcms/j_6/

ECJ 1991: European Court of Justice, Cases 63/89, Assurances du Crédit/Council and Commission, Coll. 1991, I-1799, http://curia.europa.eu/jcms/jcms/j_6/

ECJ 1992: European Court of Justice, Wallonian Waste, C-2/90, Coll. 1992, I-4431, http://curia. europa.eu/jcms/jcms/j_6/

ECJ 1993: European Court of Justice, Meyhui, C-51/93, Coll. 1994, I-3879, http://curia.europa.eu/jcms/jcms/j_6/

ECJ 1996: European Court of Justice, Francovich, C-479/93, Coll. 1996, I-3843, http://curia.europa.eu/jcms/jcms/j_6/

ECJ 1997 European Court of Justice, Kieffer und Thill, C-114/96, Coll. 1997, S. I-3629, http://curia.europa.eu/jcms/jcms/j_6/

ECJ 1997a: European Court of Justice, Centros, Coll 1997, C-212/97, I-1477, http://curia.europa.eu/jcms/jcms/j_6/

ECJ 1997b: European Court of Justice, Opinion of the Advocate General La Pergola C-212/97, I-1477, http://curia.europa.eu/jcms/jcms/j_6/

ECJ 2001: European Court of Justice, PreussenElektra, C-379/98, Coll. 2001, I-2099, http://curia.europa.eu/jcms/jcms/j_6/ and http://eur-lex.europa.eu/LexUriServ/LexUriServ.do?uri=CELEX:61998J0379:EN:HTML

ECJ 2001a: European Court of Justice, Schwarzkopf, C-169/99, Coll. 2001, I-5901, http://curia.europa.eu/jcms/jcms/j_6/

ECJ 2003: European Court of Justice, German Pharmacy Association, C-322/01, Coll. 2003, I-14887, http://curia.europa.eu/jcms/jcms/j_6/

ECJ 2003a: European Court of Justice, Bosal, C-168/01, Coll. 2003, S. I-9409, http://curia.europa.eu/jcms/jcms/j_6/

ECJ 2003b: European Court of Justice, Inspire Art, C-167/01, Coll. 2003, I-10155, http://curia.europa.eu/jcms/jcms/j_6/

ECJ 2004: European Court of Justice, Arnold André, C-434/02, Coll. 2004, I-11825, http://curia.europa.eu/jcms/jcms/j_6/

ECJ 2004a: European Court of Justice, Radlberger, C-309/02, Coll. 2004, I-11763, http://curia.europa.eu/jcms/jcms/j_6/

ECJ 2006: European Court of Justice, Keller Holding, C-471/04, Coll. 2006, S. I-2107, http://curia.europa.eu/jcms/jcms/j_6/

ECJ 2007: European Court of Justice, Diageo C-457/05, Coll. 2007, S. I-08075, http://curia.europa.eu/jcms/jcms/j_6/

ECN 2011, L.W.M. Beurskens, M. Hekkenberg, P. Vethman (Energy Research Center of the Netherlands), Renewable Energy Projections as Published in the National Renewable Energy Action Plans of the European Member States, Covering all 27 EU Member States with updates for 20 Member States, 28.11.2011, http://www.ecn.nl/docs/library/report/2010/e10069.pdf

ECOFYS 2011: Jager de, David/Klessmann, Corinna/Stricker, Eva/Winkel, Thomas/Visser de, Erika/Koper, Michèle/Ragwitz, Mario/Held, Anne/Resch, Gustav/Busch, Sebastian, Panzer, Christian/Gazzo, Alexis/Roulleau, Pierre/Henriet, Marion/Bouillé, Arnaud, Financing Renewable Energy in the European Energy Market, Ecofys, 2011, http://ec.europa.eu/energy/renewables/studies/ doc/renewables/2011_financing_renewable.pdf

EEG 2012: Act on granting priority to renewable energy sources (Renewable Energy Sources Act – EEG). Consolidated (non-binding) version of the Act in the version applicable as at 1 January 2012, Berlin, http://www.bmu.de/files/english/pdf/application/pdf/eeg_2012_en_bf.pdf

EEW 2012: Energy Efficiency Watch, Survey Report, Progress in energy efficiency policies in the EU Member States – the experts perspective, Linz 2012, http://www.energy-efficiency-watch.org/fileadmin/eew_documents/EEW2/EEW_Survey_Report.pdf

Effort Sharing Decision 2009: Official Journal of the European Union, Decision adopted jointled by the European Parliament and the Council, Decision No. 406/2009/EC of the European Parliament and of the Council of 23 April, on the effort of Member States to reduce their greenhouse gas emissions to meet the Community's greenhouse gas emission reduction commitments up to 2020, 05.06.2009, L140/136, http://eur-lex.europa.eu/LexUriServ/LexUriServ.do?uri=OJ:L:2009:140:0136:0148:EN:PDF

Electricity Directive 2001: Official Journal of the European Union, Directive 2001/77/EC of the European Parliament and of the Council of 27 September 2001 on the promotion of electricity produced from renewable energy sources in the internal electricity market, 27.10.2001,

L283/33, http://eur-lex.europa.eu/LexUriServ/LexUriServ.do?uri=OJ:L:2001:283: 0033:0033:EN: PDF

Electricity Market Directive 1996: Official Journal of of the European Communities, Directive 96/92/EC of the European Parliament and of the Council of 19 December 1996 concerning common rules for the internal market in electricity, 30.01.1997, L 27/20, http://eur-lex.europa. eu/LexUriServ/LexUriServ.do?uri=OJ:L:1997:027:0020:0029:EN:PDF

Electricity Market Directive 2003: Official Journal of the European Union, Directive 2003/54/EC of the European Parliament and of the Council of 26 June 2003 concerning common rules for the internal market in electricity and repealing Directive 96/92/EC, 15.07.2003, L176/37, http://eur-lex. europa.eu/LexUriServ/LexUriServ.do?uri=OJ:L:2003:176: 0037:0055:EN:PDF

Electricity Market Directive 2007: Proposal for a Directive of the European Parliament and of the Council amending Directive 2003/54/EC concerning common rules for the internal market in electricity (presented by the Commission), COM(2007) 528 final, Brussels 19.9.2007, http://eur-lex.europa.eu/LexUriServ/LexUriServ.do?uri=COM:2007:0528:FIN: EN:PDF

Electricity Market Directive 2009: Official Journal of the European Union, Directive 2009/72/EC of the European Parliament and of the Council of 13 July 2009 concerning common rules for the internal market in electricity and repealing Directive 2003/54/EC, 14.8.2009, L 211/55, http://eur-lex.europa.eu/LexUriServ/LexUriServ.do?uri=OJ:L:2009:211: 0055:0093:EN:PDF

Emissions Trading Directive 2003: Official Journal of the European Union, Directive2003/87/EC of the European Parliament and of the Council of 13 October 2003 establishing a scheme for greenhouse gas emission allowance trading within the Community and amending Council Directive 96/61/EC, 25.10.2003, L275/32, http://eur-lex.europa.eu/LexUriServ/LexUriServ.do? uri=OJ:L:2003:275:0032:0046:en:PDF

Emissions Trading Directive 2008: Commission of the European Communities, Proposal for a Directive of the European Parliament and of the Council amending Directive 2003/87/EC so as to improve and extend the greenhouse gas emission allowance trading system of the Community, COM(2008) 16 final, Brussels 23.1.2008, http://eur-lex.europa.eu/LexUriServ/ LexUriServ.do?uri=COM:2008:0016:FIN:en:PDF

Emissions Trading Directive 2009: Official Journal of the European Union, Directive 2009/29/EC of the European Parliament and of the Council of 23 April 2009 amending Directive 2003/87/EC so as to improve and extend the greenhouse gas emission allowance trading scheme of the Community, 05.06.2009, L140/63, http://eur-lex.europa.eu/LexUriServ/LexUriServ.do?uri=OJ:L:2009:140: 0063:0087:en:PDF

Energy Efficiency Directive 2012: Directive 2012/27/EU of the European Parliament and of the Council of 25 October 2012, on energy efficiency, amending Directives 2009/125/EC and 2010/30/EU and repealing Directives 2004/8/EC and 2006/32/EC, Official Journal of the European Union, L 315/1, 14.11.2012, http://eur-lex.europa.eu/LexUriServ/LexUriServ.do?uri=OJ:L:2012:315:0001: 0056:EN:PDF

Energy [R]evolution 2007: Greenpeace International, European Renewable Engergy Council (EREC), A Sustainable World Energy Outlook, January 2007, http://www.energyblueprint.info/fileadmin/ media/documents/energy_revolution.pdf

Energy [R]evolution 2010: Greenpeace International, European Renewable Engergy Council (EREC), energy [r]evolution, towards a fully sustainable energy supply in the EU27, Brussels, June 2010, http://www.erec.org/fileadmin/erec_docs/Documents/Publications/EU%20Energy%20[R]evolution% 20Scenario%202050.pdf

Energy Taxation Directive 2003: Official Journal of the European Union, Council Directive 2003/96/EC of 27 October 2003 restructuring the Community framework for the taxation of energy products and electricity, 31.10.2003, L 283/51, http://eur-lex.europa.eu/LexUriServ/ LexUriServ.do?uri=OJ:L: 2003:283: 0051:0070:EN:PDF

EPBD 2002: Official Journal of the European Union, Directive 2002/91/EC of the European Parliament and of the Council of 16 December 2002 on the energy performance of buildings, 4.1.2003, L1/65, http://eur-lex.europa.eu/LexUriServ/LexUriServ.do?uri=OJ:L:2003:001:0065:0071:EN:PDF

EPBD 2010: Official Journal of the European Union, Directive 2010/31/EU of the European Parliament and of the Council of 19 May 2010 on the energy performance of buildings (recast), 18.6.2010, L153/13, http://eur-lex.europa.eu/LexUriServ/LexUriServ.do?uri=OJ:L:2010:153:0013:0035:EN: PDF

EREC 2004: European Renewable Energy Council (EREC), Campaign for Take-Off, Renewable Energy for Europe (1999–2003), Sharing Skills and Achievements, Brussels 2004, http://www.erec. org/fileadmin/erec_docs/Documents/Publications/FINAL_CTO_Publication.pdf

EREC 2004a: Renewable Energy Target for Europe – 20% by 2020, Brussels 2004, http://www.erec. org/fileadmin/erec_docs/Documents/Publications/EREC_Targets_2020_def.pdf

EREC 2005: European Renewable Energy Council (EREC), Joint Declaration for a European Directive to Promote Renewable Heating and Cooling, Brussels 2005, http://www.erec.org/ fileadmin/erec_docs/Documents/Publications/EREC_RES-H.pdf

EREC 2010: European Renewable Energy Council (EREC), Renewable Energy in Europe – Markets, Trends and Technologies, Brussels, May 2010, http://www.erec.org/media/publications/erec-book.html

EREC 2010a, European Renewable Energy Council, RE-thinking 2050 – A 100% Renewable Energy Vision for the European Union, Brussels April 2010, http://www.erec.org/fileadmin/ erec_docs/Documents/Publications/ReThinking2050_full%20version_final.pdf

EREC 2011: European Renewable Energy Council, 45% by 2030 – Towards a truly sustainable energy system in the EU, Brussels, May 2011, http://www.erec.org/fileadmin/erec_docs/Documents/ Publications/45pctBy2030_ERECReport.pdf

EREC 2011a: Mapping Renewable Energy Pathways towards 2020 – EU Industry Roadmap, Brussels, http://www.erec.org/fileadmin/erec_docs/Documents/Publications/EREC-roadmap-V4_final.pdf

EREC 2012: European Renewable Energy Council, Press Release, EU Council calls for strong 2030 renewables framework, 3.12.2012, http://www.erec. org/fileadmin/erec_docs/Documents/Press_ Releases/EREC_Press_Release_-_EU_ Council_calls_for_strong_2030_renewables_framework. pdf

EREC 2013a: European Renewable Energy Council, Press Release, European Parliament calls on the Commission to assess a binding 2030 renewable energy target, 19.03.2013, http://www.erec.org/fileadmin/erec_docs/Documents/Press_Releases/EREC_PR-_RES_target_ post-2020_target.pdf

EREC 2013b: European Renewable Energy Council, Hat-trick 2030 – An integrated climate and energy framework, Brussels, April 2013, http://www.erec.org/fileadmin/erec_docs/Documents/ Publications/EREC_Hat-trick2030_April2013.pdf

ESD 2006: Official Journal of the European Union, Directive 2006/32/EC of the European Parliament and of the Council of 5 April 2006 on energy end-use efficiency and energy services and repealing Council Directive 93/76/EEC, 27.4.2006, L114/64, http://eur-lex.europa. eu/LexUriServ/LexUriServ.do?uri=OJ:L:2006:114:0064:0064:en:pdf

EUFORES 2011: REPAP2020 – Renewable Energy Policy Action Paving the Way towards 2020, Policy Conclusions And Recommendations from the National Renewable Energy Action Plans, Brussels 2011, http://www.repap2020.eu/fileadmin/user_upload/Events-docs/Brochures/ REPAP2020_PR_BrochureFINAL__2_.pdf

Eurelectric 2007: The Role of Electricity – A New Path to Secure, Competitive Energy in a Carbon-Constrained World, Brussels, March 2007, http://www2.eurelectric.org/DocShareNoFrame/ Docs/4/CLNBLPCBDBHFNAIPMAFIKAKFVHYD4QH1HLVTQLQOVQQ7/Eurelectric/docs/ DLS/Roleofelectricityfinalforwebsite-2007-030-0255-2-.pdf

Eurelectric 2009: Power Choices – Pathways to Carbon-neutral Electricity in Europe by 2050, Brussels, www.eurelectric.org/PowerChoices2050/

Eurelectric 2009a: Choices – Pathways to Carbon-neutral Electricity in Europe by 2050, Lars G. Josefsson, President of Eurelectric, Power choices study launch event European Parliament, 10 November 2009, http://www.eurelectric.org/media/43863/final_presentation_lars_g_josefsson_-_ for_the_website-2009-030-0946-01-e.pdf

European Commission 1996: European Commission, Communication from the Commission, Energy for the Future: Renewable Sources of Energy, Green Paper for a Community Strategy, COM (96) 576 final, Brussels, 20.11.1996, http://eur-lex.europa.eu/LexUriServ/LexUriServ.do?uri=COM:1996:0576:FIN:EN:PDF

European Commission 1997a: European Commission, Communication from the Commission to the European Parliament, the Council, the European Economic and Social Committee and the Committee of the Regions, The Energy Dimension of Climate Change, COM(97) 196 final, Brussels, 14.5.1997, http://aei.pitt.edu/4723/1/000817_1.pdf

European Commission 1997b: European Commission, Communication from the Commission to the European Parliament, the Council, the European Economic and Social Committee and the Committee of the Regions, Climate Change – The EU Approach for Kyoto, COM(97) 481 final, Brussels, 1.10.1997, http://aei.pitt.edu/6244/1/6244.pdf

European Commission 1997c: European Commission, Communication from the Commission, Energy for the Future, Renewable Sources of Energy, White Paper for a Community Strategy and Action Plan, COM(97) 599 final, Brussels, 26.11.1997, http://europa.eu/documents/comm/white_papers/pdf/com97_599_en.pdf

European Commission 2001: Communication from the Commission to the European Parliament, the Council, the European Economic and Social Committee and the Committee of the Regions on the implementation of the Community Strategy and Action Plan on Renewable Energy Sources (1998–2000), COM(2001) 69 final, Brussels, 16.2.2001, http://eur-lex.europa.eu/LexUriServ/LexUriServ.do?uri=COM:2001:0069:FIN:EN:PDF

European Commission 2005: Commission of the European Communities, Communication from the Commission, The support of electricity from renewable energy sources, COM (2005) 627 final, Brussels, 7.12.2005, http://eur-lex.europa.eu/LexUriServ/LexUriServ.do?uri=COM:2005:0627:FIN:EN:PDF

European Commission 2005a: Communication from the Commission to the Council and the European Parliament, Report on progress in creating the internal gas and electricity market, COM (2005) 568 final, Brussels, 15.11.2005, http://ec.europa.eu/energy/electricity/report_2005/doc/2005_report_en.pdf

European Commission 2006: Press Release, The Commission to act over EU energy markets, MEMO/06/481, 12.12.2006, http://europa.eu/rapid/press-release_MEMO-06-481_en.htm

European Commission 2007: Communication from the Commission to the Council and the European Parliament, Renewable Energy Road Map, Renewable energies in the 21st century: building a more sustainable future, COM(2006) 848 final, Brussels, 10.1.2007, http://eur-lex.europa.eu/LexUriServ/LexUriServ.do?uri=COM:2006:0848:FIN:EN:PDF

European Commission 2007a: Communication from the Commission to the Council and the European Parliament, Prospects for the internal gas and electricity market, COM(2006) 841 final, Brussels 10.01.2007, http://eur-lex.europa.eu/LexUriServ/site/en/com/2006/com2006_0841en01.pdf

European Commission 2007b: Communication from the Commission: Inquiry pursuant to Article 17 of Regulation (EC) No 1/2003 into the European gas and electricity sectors (Final Report), COM(2006) 851 final, Brussels 10.01.2007, http://eur-lex.europa.eu/LexUriServ/LexUriServ.do?uri=COM:2006:0851:FIN:EN:PDF

European Commission 2007c: Communication from the Commission to the Council and the European Parliament, An Energy Policy for Europe, COM(2007) 1 final, http://ec.europa.eu/energy/energy_policy/doc/01_energy_policy_for_europe_en.pdf

European Commission 2008: Commission of the European Communities, Commission Staff Working Document, The support of electricity from renewable sources, Accompanying document to the Proposal for a Directive of the European Parliament and of the Council on the promotion of the use of energy from renewable sources, SEC (2008) 57, Brussels, 23.01.2008, http://ec.europa.eu/energy/climate_actions/doc/2008_res_working_document_en.pdf

European Commission 2008a: Commission of the European Communities, Proposal for a Directive of the European Parliament and of the Council on the promotion of the use of energy from renewable sources, COM (2008) 19 final, Brussels, 23.1.2008, http://ec.europa.eu/energy/climate_actions/doc/2008_res_directive_en.pdf

European Commission 2010: Communication from the Commission to the European Parliament, the Council, the European Economic and Social Committee and the Committee of the Regions, Energy 2020 – A strategy for competitive, sustainable and secure energy, COM (2010) 639 final, Brussels, 10.11.2010, http://eur-lex.europa.eu/LexUriServ/LexUriServ.do?uri=COM:2010:0639:FIN:EN:PDF

European Commission 2010a: Communication from the Commission, Europe 2020 A strategy for smart, sustainable and inclusive growth, COM (2010) 2020 final, Brussels, 3.3.2010, http://eur-lex.europa.eu/LexUriServ/LexUriServ.do?uri=COM:2010:2020:FIN:EN:PDF

European Commission 2011: The Internal Energy Market – Time to Switch into a Higher Gear, Non-paper, downloaded from Energy Website of the European Commission on 24 February 2013: http://ec.europa.eu/energy/gas_electricity/legislation/doc/20110224_non_paper_internal_nergy_market.pdf

European Commission 2011a: Communication from the Commission to the European Parliament, the Council, the European Economic and Social Committee and the Committee of the Regions, A Roadmap for moving towards a competitive low carbon economy in 2050, COM(2011) 112 final, Brussels, 8.3.2011, http://eur-lex.europa.eu/LexUriServ/LexUriServ.do?uri=COM:2011:0112:FIN:EN:PDF

European Commission 2011b: Commission Staff Working Document, Impact Assessment, Accompanying document to the Communication from the Commission to the European Parliament, the Council, the European Economic and Social Committee and the Committee of the Regions, A Roadmap for moving towards a competitive low carbon economy in 2050, SEC(2011) 288 final, Brussels, 8.3.2011, http://eur-lex.europa.eu/LexUriServ/LexUriServ.do?uri=SEC:2011:0288:FIN:EN:PDF

European Commission 2011c: Commission Staff Working Document, Summary of the Impact Assessment, Accompanying document to the Communication from the Commission to the European Parliament, the Council, the European Economic and Social Committee and the Committee of the Regions, A Roadmap for moving towards a competitive low carbon economy in 2050, SEC(2011) 289 final, Brussels, 8.3.2011, http://eur-lex.europa.eu/LexUriServ/LexUriServ.do?uri=SEC:2011:0289:FIN: EN:PDF

European Commission 2011d: White Paper, Roadmap to a Single European Transport Area – Towards a competitive and resource efficient transport system, COM (2011) 144 final, Brussels, 28.3.2011, http://eur-lex.europa.eu/LexUriServ/LexUriServ.do?uri=COM:2011:0144:FIN:en:PDF

European Commission 2011e: Communication from the Commission to the European Parliament, the Council, the European Economic and Social Committee and the Committee of the Regions, On security of energy supply and international cooperation – "The EU Energy Policy: Engaging with Partners beyond Our Borders", COM(2011) 539 final, Brussels, 7.9.2011, http://eur-lex.europa.eu/LexUriServ/LexUriServ.do?uri=COM:2011:0539:FIN:EN:PDF

European Commission 2011f: Communication from the Commission to the European Parliament, the Council, the European Economic and Social Committee and the Committee of the Regions, Energy Roadmap 2050, COM(2011) 885 final, Brussels, 15.12.2011, http://eur-lex.europa.eu/LexUriServ/LexUriServ.do?uri=COM:2011:0885:FIN: EN:PDF

European Commission 2011g: Commission Staff Working Paper, Executive Summary of the Impact Assessment accompanying the document Communication from the Commission to the European Parliament, the Council, the European Economic and Social Committee and the Committee of the Regions, Energy Roadmap 2050, SEC(2011) 1566 final, Brussels, 15.12.2011, http://eur-lex.europa.eu/LexUriServ/LexUriServ.do?uri=SEC:2011:1566:FIN: EN:PDF

European Commission 2011h: Commission Staff Working Paper, Impact Assessment accompanying the document Communication from the Commission to the European Parliament, the Council, the European Economic and Social Committee and the Committee of the Regions, Energy Roadmap 2050, SEC(2011) 1565/2, Brussels, http://ec.europa.eu/energy/energy2020/roadmap/doc/sec_2011_1565_part1.pdf and http://ec.europa.eu/energy/energy2020/roadmap/doc/sec_2011_1565_part2.pdf

European Commission 2011i: Communication from the Commission to the European Parliament and the Council, Renewable Energy: Progressing towards the 2020 target, COM (2011) 31 final, Brussels, 31.01.2011, http://eur-lex.europa.eu/LexUriServ/LexUriServ.do?uri=COM:2011:0031:FIN:EN:PDF

European Commission 2011j: Commission of the European Communities, Commission Staff Working Document, Recent progress in developing renewable energy sources and technical evaluation of the use of biofuels and other renewable fuels in transport in accordance with Article 3 of Directive 2001/77/EC and Article 4(2) of Directive 2003/30/EC, Accompanying Document to the Communication from the Commission to the European Parliament and the Council, Renewable Energy: Progressing towards the 2020 target, SEC (2011) 130 final, Brussels, 31.01.2011, http://ec.europa.eu/energy/renewables/reports/doc/sec_2011_0130.pdf

European Commission 2011k: Commission of the European Communities, Commission Staff Working Document, Review of European and national financing of renewable energy in accordance with Article 23(7) of Directive 2009/28/EC, Accompanying Document to the Communication from the Commission to the European Parliament and the Council Renewable Energy: Progressing towards the 2020 target, SEC(2011) 131 final, Brussels, 31.1.2011, http://eur-lex.europa.eu/LexUriServ/LexUriServ.do?uri=SEC:2011:0131:FIN:EN:PDF

European Commission 2011l: Proposal for a Regulation of the European Parliament and of the Council amending Council Regulation (EC) No 1217/2009 setting up a network for the collection of accountancy data on the incomes and business operation of agricultural holdings in the European Community, COM (2011) 855 final, Brussels, 7.12.2011, http://eur-lex.europa.eu/LexUriServ/LexUriServ.do?uri=COM:2011:0855:FIN:EN:PDF

European Commission 2011m: European Commission, Public Consultation on the Renewable Energy Strategy, Consultation document, Brussels, 6.12.2011, http://ec.europa.eu/energy/renewables/consultations/doc/20120207_renewable_energy_strategy.pdf

European Commission 2012a: Communication from the Commission to the European Parliament, the Council, the European Economic and Social Committee and the Committee of the Regions, EU State Aid Modernisation (SAM), COM(2012) 209 final, Brussels, 8.5.2012, http://eur-lex.europa.eu/LexUriServ/LexUriServ.do?uri=COM:2012:0209:FIN: EN:PDF

European Commission 2012b: Communication from the Commission to the European Parliament, the Council, the European Economic and Social Committee and the Committee of the Regions, Renewable Energy: a major player in the European energy market, COM (2012) 271 final, Brussels, 6.6.2012, http://ec.europa.eu/energy/renewables/doc/communication/2012/comm_de. pdf

European Commission 2012c: Commission Staff Working Paper, Impact Assessement, Accompanying the document Communication from the Commission to the European Parliament, the Council, the European Economic and Social Committee and the Committee of the Regions, Renewable Energy: a major player in the European energy market, SWD(2012) 149 final, Brussels, 6.6.2012, http://eur-lex.europa.eu/LexUriServ/LexUriServ.do?uri=SWD:2012:0149:FIN:EN:PDF

European Commission 2012d: Commission Staff Working Paper, Executive Summary of the Impact Assessement, Accompanying the document Communication from the Commission to the European Parliament, the Council, the European Economic and Social Committee and the Committee of the Regions, Renewable Energy: a major player in the European energy market, SWD(2012) 163 final, Brussels, 6.6.2012, http://eur-lex.europa.eu/LexUriServ/LexUriServ.do?uri=SWD:2012:0163:FIN: EN:PDF

European Commission 2012e: Commission Staff Working Document, Accompanying the document Communication from the Commission to the European Parliament, the Council, the European Economic and Social Committee and the Committee of the Regions, Renewable Energy: a major player in the European energy market, SWD (2012) 164 final, 6.6.2012, http://eur-lex.europa.eu/LexUriServ/LexUriServ.do?uri=SWD:2012:0164:FIN:EN:PDF

European Commission 2012f: Proposal for a Directive of the European Parliament and of the Council amending Directive 98/70/EC relating to the quality of petrol and diesel fuels and amending Directive 2009/28/EC on the promotion of the use of energy from renewable sources,

COM (2012) 595 final, Brussels, 17.10.2012, http://ec.europa.eu/energy/renewables/biofuels/doc/biofuels/com_2012_0595_en.pdf

European Commission 2012g: Commission Staff Working Document, Impact Assessment, Accompanying the document Proposal for a Directive of the European Parliament and of the Council amending Directive 98/70/EC relating to the quality of petrol and diesel fuels and amending Directive 2009/28/EC on the promotion of the use of energy from renewable sources, SWD(2012) 343 final, Brussels, 17.10.2012, http://ec.europa.eu/energy/renewables/biofuels/doc/biofuels/swd_2012_0343_ia_en.pdf

European Commission 2012h: Commission Staff Working Document, Executive Summary of the Impact Assessemt on indirect land-use change related to biofuels and bioliquids, Accompanying the document Proposal for a Directive of the European Parliament and of the Council amending Directive 98/70/EC relating to the quality of petrol and diesel fuels and amending Directive 2009/28/EC on the promotion of the use of energy from renewable sources, SWD(2012) 344 final, Brussels, 17.10.2012, http://ec.europa.eu/energy/renewables/biofuels/doc/biofuels/swd_2012_0344_ia_resume_en.pdf

European Commission 2012i: Commission Regulation COM(2012) .../.. of XXX amending Regulation (EU) No 1031/2010 in particular to determine the volumes of greenhouse gas emission allowances to be auctioned in 2013-2020, Brussels, 12.11.2012 draft, http://ec.europa.eu/clima/policies/ets/cap/auctioning/docs/20121112_com_en.pdf

European Commission 2012j: Commission Staff Working Document, Proportionate Impact Assessment accompanying the document Commission Regulation COM (2012) .../.. of XXX amending Regulation (EU) No 1031/2010 in particular to determine the volumes of greenhouse gas emission allowances to be auctioned in 2013-2020, Brussels, 12.11.2012 draft, http://ec.europa.eu/clima/policies/ets/cap/auctioning/docs/20121112_swd_en.pdf

European Commission 2012k: Commission Staff Working Document, Executive Summary of the Impact Assessment accompanying the document Commission Regulation COM(2012) .../.. of XXX amending Regulation (EU) No 1031/2010 in particular to determine the volumes of greenhouse gas emission allowances to be auctioned in 2013-2020, Brussels, 12.11.2012 draft, http://ec.europa.eu/clima/policies/ets/cap/auctioning/docs/20121112_com_en.pdf

European Commission 2012l: Communication from the Commission to the European Parliament, the Council, the European Economic and Social Committee and the Committee of the Regions, Singe Market Act II – Together for new growth, COM(2012) 573 final, Brussels, 3.10.2012, http://ec.europa.eu/internal_market/smact/docs/single-market- act2_en.pdf

European Commission 2012m: Communication from the Commission to the European Parliament, the Council, the European Economic and Social Committee and the Committee of the Regions, Making the internal energy market work, COM(2012) 663 final, Brussels, 15.11.2012, http://eur-lex.europa.eu/LexUriServ/LexUriServ.do?uri=COM:2012:0663:FIN: EN:PDF

European Commission 2012n: Commission Staff Working Document, Investment projects in energy infrastructure, accompanying the document Communication from the Commission to the European Parliament, the Council, the European Economic and Social Committee and the Committee of the Regions, Making the internal energy market work, SWD(2012) 367 final, 15.11.2012, http://ec.europa.eu/energy/gas_electricity/doc/20121115_iem_swd_0367_en.pdf

European Commission 2012o: Commission Staff Working Document, Energy Markets in the European Union in 2011, accompanying the document Communication from the Commission to the European Parliament, the Council, the European Economic and Social Committee and the Committee of the Regions, Making the internal energy market work, SWD (2012) 368 final, Parts I, II, III, 15.11.2012, published together in a brochure to be downloaded at http://ec.europa.eu/energy/gas_electricity/doc/20121217_energy_market_2011_lr_en.pdf

European Commission 2013: Environmental and Energy Aid Guidelines 2014-2010, Consultation Paper, Brussels, 11 March 2013, http://ec.europa.eu/competition/state_aid/legislation/environmental_aid_issues_paper_en.pdf

European Commission 2013a: Green Paper, A 2030 framework for climate and energy policies. COM (2013) 169 final, Brussels, 27.3.2013, http://ec.europa.eu/energy/consultations/doc/com_2013_0169_green_paper_2030_en.pdf

European Commission 2013b: Press Release, Commission moves forward on climate and energy towards 2030, Brussels, 27.3.20132, http://europa.eu/rapid/press-release_IP-13-272_en.htm

European Commission 2013c: Report from the Commission to the European Parliament, the Council, the Euroepan Social and Economic Committee and the Committee of the Regions, Renewable energy progress report, COM(2013) 175 final, Brussels, 27.3.2013, http://ec.europa.eu/energy/renewables/reports/doc/com_2013_0175_res_en.pdf

European Commission 2013d: Communication from the Commission to the European Parliament, the Council, the European Social and Economic Committee and the Committee of the Regions on the Future of Carbon Capture and Storage in Europe, COM(2013) 180 final , Brussels, 27.3.2013, http://ec.europa.eu/energy/coal/doc/com_2013_0180_ccs_en.pdf

European Commission 2013e: Consultative Communication on the future of Carbon Capture and Storage in Europe, MEMO/13/276, Brussels, 27 March 2013, http://europa.eu/rapid/press-release_MEMO-13-276_en.htm

European Parliament 2007: European Parliament resolution of 10 July 2007 on prospects for the internal gas and electricity market (2007/2089(INI)), P6_TA(2007)0326, http://www.europarl.europa.eu/sides/getDoc.do?pubRef=-//EP//NONSGML+TA+P6-TA-2007-0326+0+DOC+PDF+V0//EN

European Parliament 2008: Press Release 13.12.2008, MEPs and Council Presidency reach deal on final details of climate package, Brussels, http://www.europarl.europa.eu/sides/getDoc.do?pubRef=-//EP//TEXT+IM-PRESS+20081209IPR44022+0+DOC+XML+V0//EN&language=EN

European Parliament 2012: European Parliament, Committee on Industry, Research and Energy, Draft Report, Current challenges and opportunities for renewable energy on the European energy market, 2012/2259(INI), Rapporteur: Herbert Reul, 9.11.2012, http://www.europarl.europa.eu/sides/getDoc.do?pubRef=-%2F%2FEP%2F%2FNONSGML%2BCOMPARL%2BPE-497.809%2B01%2BDOC%2BPDF%2BV0%2F%2FEN

European Parliament 2013: Report Current challenges and opportunities for renewable energy on the European energy market, 2012/2259 (INI), Committee on Industry, Research and Energy, Rapporteur: Herbert Reul, A7-0135/2013, 28.03.2013, http://www.europarl.europa.eu/sides/getDoc.do?pubRef=-//EP//NONSGML+REPORT+A7-2013-0135+0+ DOC+PDF+V0//EN

European Parliament 2013a: European Parliament resolution of 21 May 2013 current challenges and opportunities for renewable energy in the European internal energy market, European Parliament resolution of 21 May 2013, P7_TA(2013)0201, http://www.europarl.europa.eu/sides/getDoc.do?pubRef=-//EP//TEXT+TA+P7- TA-2013-0201+0+DOC+XML+V0//EN&language=EN

EWEA 2012: European Wind Energy Associationn, Creating the Internal Energy Market in Europe, Brussels, September 2012, http://www.ewea.org/uploads/tx_err/Internal_energy_market.pdf

EWI 2010: Fürsch, Michaela/Golling, Christiane/Nicolosi, Marco/Wissen, Ralf/Lindenberger, Dietmar, European RES-E Policy Analysis, 2010, http://www.ewi.uni-koeln.de/fileadmin/user_upload/Publikationen/Studien/Politik_und_Gesellschaft/2010/EWI_2010-04-26_RES-E-Studie_Teil1.pdf, http://www.ewi.uni-koeln.de/fileadmin/user_upload/Publikationen/Studien/Politik_und_Gesellschaft/2010/EWI_2010-04-26_RES-E-Studie_Teil2.pdf

FHI/EEG 2011: Mario Ragwitz, Gustav Resch, Sebastian Busch, Anne Held, Daniel Rosende, Florian Rudolf, Gerda Schubert, Simone Steinhilber, Fraunhofer ISI, Karlsruhe and Energy Economics Group TU Vienna, prepared for the project Renewable Energy Policy Action Paving the Way towards 2020, Assessment of National Renewable Energy Action Plans, Augsut 2011, http://www.repap2020.eu/fileadmin/user_upload/Roadmaps/D115-Assessment_of_NREAPs__REPAP_report_-_final_edition_.pdf

Fischer/Geden 2012: Severin Fischer/Oliver Geden, Die "Energy Roadmap 2050" der EU: Ziele ohne Steuerung, SWP-Aktuell 8/2012, pp. 1–4, http://www.swp-berlin.org/fileadmin/contents/products/aktuell/2012A08_fis_gdn.pdf

Fuel Quality Directive 2009: Official Journal of the European Union, Directive 2009/30/EC of the European Parliament and of the Council of 23 April 2009 amending Directive 98/70/EC as regards the specification of petrol, diesel and gas-oil and introducing a mechanism to monitor and reduce greenhouse gas emissions and amending Council Directive 1999/32/EC as regards the specification of fuel used by inland waterway vessels and repealing Directive 93/12/EEC, 5.6.2009, L140/88, http://eur-lex.europa.eu/LexUriServ/LexUriServ.do?uri=OJ:L:2009:140:0088:0113:EN:PDF

Futures-e 2008: Vienna University of Technology, Energy Economics Group (EEG), Austria in cooperation with Fraunhofer Institute Systems and Innovation Research, Karlsruhe, Germany, within the scope of the futures-e project, futures-e – 20% RES by 2020 – a balanced scenario to meet Europes RE target, http://www.futures-e.org/

Futures-e 2009: Vienna University of Technology, Energy Economics Group (EEG), Austria in cooperation with Fraunhofer Institute Systems and Innovation Research, Karlsruhe, Germany, within the scope of the futures-e project, Scenarios on future European polices for Renewable Electricity – 20% RES by 2020, http://www.futures-e.org/

Gas Market Directive 1998: Official Journal of of the European Communities, Directive 98/30/EC of the European Parliament and of the Council of 22 June 1998 concerning common rules for the internal market in natural gas, 21.07.1998, L 204/2, http://eur-lex.europa.eu/LexUriServ/LexUriServ.do?uri=OJ:L:1998:204:0001:0012:EN:PDF

Gas Market Directive 2003: Official Journal of the European Union, Directive 2003/55/EC of the European Parliament and of the Council of 26 June 2003 concerning common rules for the internal market in natural gas and repealing Directive 98/30/EC, 15.07.2003, L 176/57, http://eur-lex.europa.eu/LexUriServ/LexUriServ.do?uri=OJ:L:2003:176:0057:0057:EN:PDF

Gas Market Directive 2007: Proposal for a Directive of the European Parliament and of the Council amending Directive 2003/55/EC concerning common rules for the internal market in natural gas, (presented by the Commission), COM(2007) 529 final, Brussels 19.9.2007, http://eur-lex.europa.eu/LexUriServ/LexUriServ.do?uri=COM:2007:0529:FIN:EN:PDF

Gas Market Directive 2009: Official Journal of the European Union, Directive 2009/73/EC of the European Parliament and of the Council of 13 July 2009 concerning common rules for the internal market in natural gas and repealing Directive 2003/55/EC, 14.8.2009, L 211/94, http://eur-lex.europa.eu/LexUriServ/LexUriServ.do?uri=OJ:L:2009:211:0094:0136:en:PDF

Gilbertson and Reyes 2009: Tamra Gilbertson and Oscar Reyes, Carbon Trade Watch, Carbon Trading – How it works and why it fails, Dag Hammarskjöld Foundation, Uppsala 2009, http://webcache.googleusercontent.com/search?q=cache:http://www.dhf.uu.se/pdffiler/cc7/cc7_web_low.pdf

GWEC 2013: Global Wind Energy Council (GWEC), Global Wind Statistics 2012, Brussels, 11.2.2013, http://www.gwec.net/wp-content/uploads/2013/02/GWEC-PRstats-2012_english.pdf

GSR 2012: REN21, Renewables 2011 Global Status Report, Paris, June 2012, http://www.ren21.net/Portals/97/documents/GSR/REN21_GSR2011.pdf

GSR 2013: REN21, Renewables 2012 Global Status Report, Paris, to be published in June 2013, www.ren21.net

Hayek 1968: F. A. von Hayek, Der Wettbewerb als Entdeckungsverfahren, lecture, held on 05.07.1968 at the Institute for the World Economy at Kiel University, printed in: von Hayek, F.A., Freiburger Studien, 2nd edition, Tübingen 1994, p. 249 ff.

IEA 2011: International Energy Agency, Harnessing Variable Renewables – A Guide to the Balancing Challenge, OECD/IEA 2011, http://www.oecd-ilibrary.org/energy/harnessing-variable-renewables_9789264111394-en

IEA 2012: International Energy Agency, Renewable Energy Medium-Term Market Report 2012, Martket Trends and Projections to 2017, OECD/IEA 2012, http://www.iea.org/w/bookshop/add.aspx?id=432

IEA-RETD 2012: Rolf de Vos, Janet Sawin, READy, Renewable Energy Action on Deployment, Presenting: The ACTION Star; six policy ingredients for accelerated deployment of renewable energy, Elsevier 2012, http://www.elsevier.com/books/ready-renewable-energy-action-on-deployment/iea-retd/978-0-12-405519-3#

IFIC 2010: Mario Ragwitz, Anne Held (Fraunhofer ISI), Eva Stricker, Anja Krechting (Ecofys), Gustav Resch, Christian Panzer (EEG), Recent experiences with feed-in tariff systems in the

EU – A research paper for the International Feed-In Cooperation (IFIC), November 2010, A report commissioned by the [German] Minstry for the Environment, Nature Conservation and Nuclear Safety (BMU), http://www.futurepolicy.org/fileadmin/user_upload/PDF/Feed_in_Tariff/IFIC_feed-in_evaluation_Nov_2010.pdf

IMF 2007: International Monetary Fund, World Economic and Financial Surveys, World Economic Outlook, October 2007, Globalisation and Inequality, http://www.imf.org/external/pubs/ft/weo/2007/02/pdf/text.pdf

IPCC 2012: IPCC Special Report on Renewable Energy Sources and Climate Change Mitigation. Prepared by Working Group III of the Intergovernmental Panel on Climate Change [O. Edenhofer, R. Pichs-Madruga, Y. Sokona, K. Seyboth, P. Matschoss, S. Kadner, T. Zwickel, P. Eickemeier, G. Hansen, S. Schlömer, C. von Stechow (eds)]. Cambridge University Press, Cambridge, United Kingdom and New York, NY, USA, published 2012, http://srren.ipcc-wg3.de/report/IPCC_SRREN_Full_Report.pdf

Klessmann 2011: Corinna Kleßmann, Increasing the effectiveness and efficiency of renewable energy support policies in the European Union, 2011, http://igitur-archive.library.uu.nl/dissertations/2011-1222-200420/klessmann.pdf

Lange 2013: Bernd Lange, MEP, Pressemitteilung, Europa muss mehr für Erneuerbare tun, Brüssel, 19.3.2013, http://www.bernd-lange.de/aktuell/nachrichten/2013/362842.php

Legislation Summary: Internal market for energy (until March 2011), viewed on 22 February 2013, http://europa. eu/legislation_summaries/energy/internal_energy_market/l27005_en.htm

Lisbon Treaty: Official Journal of the European Union, Consolidated Versions of the Treaty on European Union and the Treaty on the Functioning of the European Union (2010/C 83/01), 30.03.2010, C83, http://eur-lex.europa.eu/LexUriServ/LexUriServ.do?uri=OJ:C:2010:083:FULL:EN:PDF

Matthes 2012: Felix Chr. Matthes, Langfristperspektiven der europäischen Energiepolitik – Die Energy Roadmap 2050 der Europäischen Union, ET 2012, Book 1/2, pp. 50–53, http://www.et-energie-online.de/Zukunftsfragen/tabid/63/NewsId/22/Langfristperspektiven-der-europaischen-Energiepolitik–Die-Energy-Roadmap-2050-der-Europaischen-Union.aspx?error=Die%20Eingabezeichenfolge%20hat%20das%20falsche%20Format.

Matthes 2013: Felix Chr. Matthes, Öko-Institut e.V., The European Emission Trading System: current state and way forward, Presentation at European Federation of Local Energy Companies (CEDEC) Congress "Connecting Customers and Climate: How local energy companies invest and facilitate the Future", Brussels, 19.3.2013, http://www.cedec.com/files/default/2013-03-19-matthes-oeko-institut. pdf

May 2012: Hanne May, Oettingers Visionen, neue energie 2012, Book 1, pp. 16–18.

Mendonca 2007: Miguel Mendonca, In Tariffs: Accelerating the Deployment of Renewable Energy, London, Earthscan 2007.

Müller 2009: Thorsten Müller, Neujustierung des europäischen Umweltenergierechts im Bereich Erneuerbarer Energien? – Zur Richtlinie zur Förderung der Nutzung von Energie aus erneuerbaren Quellen, in: Cremer, Wolfram/Pielow, Johann-Christian (Hrsg.), Probleme und Perspektiven im Energieumweltrecht, Stuttgart 2009, pp. 143–175, http://www.stiftung-umweltenergierecht. de/forschung/ mitarbeiter/thorsten-mueller/wissenschaftliche- veroeffentlichungen.html

OPTRES 2007: Ragwitz, Mario/Held, Anne/Resch, Gustav/Faber, Thomas/Haas Reinhard/Huber, Claus/Morthorst, Poul Erik/Jense, Stine Grenaa/Coenraads Rogier/Voogt, Monique/Reece, Gemma/Konstantinaviciute, Inga/Heyder, Bernhard, OPTRES Final Report, Assessment and optimization of renewable energy support schemes in the European electricity market, 2007, http://ec.europa.eu/energy/renewables/studies/doc/renewables/2007_02_optres_recommendations.pdf

platts 2011: platts Renewable Energies Report, Issue 234, July, 2011, http://www.platts.com/IM.Platts. Content/ProductsServices/Products/renewableenergyreport.pdf

Posner 2007: Richard A. Posner, Economic Analysis of Law, Aspen Publications 2007.

PRIMES 2010: National Technical University of Athens, E3M-Lab for the European Commission, Directorate for Energy, EU energy trends to 2030, Update 2009, Brussels/Athens, August 2010, http://www.e3mlab.ntua.gr/DEFAULT.HTM

PWC 2010: 100% Renewable Electricity – A Roadmap to 2050 for Europe and North Africa, London 2010, http://www.pwc.co.uk/en_UK/uk/assets/pdf/100-percent-renewable-electricity.pdf

Renewables Directive 2008: Commission of the European Communities, Draft Proposal for a Directive on the promotion of the use of energy from renewable sources, COM (2008) 19 final, Brussels, 23.01.2008, http://ec.europa.eu/energy/climate_actions/doc/2008_res_directive_en.pdf

Renewables Directive 2009: Official Jounal of the European Union, Directive 2009/28/EC of the European Parliament and of the Council of 23 April 2009 on the promotion of the use of energy from renewable sources and amending and subsequently repealing Directives 2001/77/EC and 2003/30/EC, 05.06.2009, L140/16, http://eur-lex.europa.eu/LexUriServ/ LexUriServ.do?uri=OJ:L:2009:140: 0016:0062:en:PDF

Reich 1992: Norbert Reich, Competition between legal orders: A new paradigm of EC law? CMLRev 1992, 861.

Regulation 1/2003: Official Journal of the European Communities, Council Regulation (EC) 1/2003 of 16 December 2002 on the implementation of the rules on competition laid down in Articles 81 and 82 of the Treaty, 4.1.2003, L 1/1, http://eur-lex.europa.eu/LexUriServ/ LexUriServ.do?uri=OJ:L:2003:001: 0001:0025:EN:PDF

Regulation 1228/2003: Official Journal of the European Union, Regulation (EC) of the European Parliament and of the Council of 26 June 2003 on conditions for access to the network for cross-border exchanges in electricity, 15.07.2003, L 176/1, http://eur-lex.europa.eu/ LexUriServ/LexUriServ.do?uri=OJ:L:2003:176: 0001:0010:EN:PDF

Regulation 713/2009: Official Journal of the European Union, Regulation (EC) No 713/2009 of the European Parliament and of the Council of 13 July 2009 establishing an Agency for the Cooperation of Energy Regulators, 14.8.2009, L 211/1, http://eur-lex.europa.eu/LexUriServ/ LexUriServ.do?uri=OJ:L:2009:211: 0001:0014:EN:PDF

Regulation 714/2009: Official Journal of the European Union, Regulation (EC) No 714/2009 of the European Parliament and of the Council of 13 July 2009 on conditions for access to the network for cross-border exchanges in electricity and repealing Regulation (EC) No 1228/2003, 14.8.2009, L 211/15, http://eur-lex.europa.eu/LexUriServ/LexUriServ.do?uri=OJ:L:2009:211:0015:0035:EN: PDF

Regulation 715/2009: Official Journal of the European Union, Regulation (EC) No 715/2009 of the European Parliament and of the Council of 13 July 2009 on conditions for access to the natural gas transmission networks and repealing Regulation (EC) No 1775/2005, 14.8.2009, L 211/36, http://eur-lex.europa.eu/LexUriServ/LexUriServ.do?uri=OJ:L:2009:211:0036:0054:en:PDF

REShaping 2010: Gustav Resch (EEG), Mario Ragwitz (Fraunhofer ISI), Quo(ta) vadis, Europe? A comparative assessment of two recent studies on the future development of renewable electricity support in Europe (EWI and futures-e), A report compiled within the European research project RE-Shaping, Vienna, November 2010, http://www.reshaping-res-policy.eu/downloads/Quo(ta)-vadis-Europe_RE-Shaping-report.pdf

REShaping 2010a: Kleßmann, Corinna/ Lamers, Patrick/ Ragwitz, Mario/ Resch, Gustav, Design options for cooperation mechanisms under the new European renewable energy directive, Energy Policy 2010, pp. 4679–4691, http://www.reshaping-res-policy.eu/downloads/D4_report_design-options-RES-flexibility-mechanisms.pdf

RES4less 2011: RES4less newsletter, 2nd Issue, August 2011, http://www.res4less.eu/files/nl/nl2/ RES4LESS_2nd_newsletter.pdf

Reul 2013: Herbert Reul, MEP, Pressemitteilung, Erneuerbare Energien: Mehr Europa notwendig – Einheitliche Förderkriterien notwendig/Nationale Fördersysteme verzerren Binnenmarkt/Gegen Verdopplung der verpflichtenden EU-Ausbauziele, Brüssel, 19.3.2013, http://www.herbert-reul.de/ index.php?id = 14 & tx _ttnews % 5Byear%5D =2013&tx _ ttnews%5Bmonth%5D=03&tx_ttnews% 5Bday%5D=19&tx_ttnews%5Btt_news%5D=1509&cHash=79865643dcb0aa546bca39cb01339b89

REVE 2012: Revista Eólica y del Vehículu Eléctrico (reve), Spain's conservative government decreed a moratorium on renewable energy, 1.12.2012, http://www.evwind.es/2012/02/01/spains-conservative-government-decreed-a-moratorium-on-renewable-energy/16325, viewed 26 April 2013.

Streit 1996: Manfred E. Streit, Competition among systems, Harmonisation and European Integration, Max Plank Institut zur Erforschung von Wirtschaftssystemen, 1996.

TFEU 2012: Official Journal of the European Union, Consolidated version of the Treaty on the Functioning of the European Union (2012/C 326/01), 26.10.2012, C326, http://eur-lex. europa.eu/LexUriServ/LexUriServ.do?uri=OJ:C:2012:326:FULL:EN:PDF

Welt 2013: Stefanie Bolzen, Brüssel will EEG ein Ende machen – Parlament schlägt EU-weite Förderung vor, Welt digital 9.11.12 (viewed 7.4.13), http://www.welt.de/print/die_welt/wirtschaft/ article110830588/Bruessel-will-EEG-ein-Ende-machen.html

WEO 2009: International Energy Agency, World Energy Outlook 2009, Paris, November 2009, http://www.oecd-ilibrary.org/energy/world-energy-outlook-2009_weo-2009-en

WEO 2010: International Energy Agency, World Energy Outlook 2011, Paris, November 2010, http://www.iea.org/publications/freepublications/publication/weo2010.pdf

WEO 2011: International Energy Agency, World Energy Outlook 2011, Paris, November 2011, http://www.iea.org/W/bookshop/add.aspx?id=433

WEO 2012: International Energy Agency, World ENergy Outlook 2012, Paris November 2012, http://www.iea.org/W/bookshop/add.aspx?id=433%20

WWEA 2012: The World Wind Energy Association (WWEA), 2012 Half Year Report, Bonn, October 2013, http://www.wwindea.org/webimages/Half-year_report_2012.pdf

WWF 2011: WWF, ECOFYS, OMA, The Energy Report, 100% Renewable Energy by 2050, Gland/Switzerland, http://wwf.panda. org/what_we_do/footprint/climate_carbon_energy/energy_ solutions/renewable_energy/sustainable_energy_report/

WWF 2013: WWF European Policy Office, Re-energising Europe – Putting Europe on Track for 100% Renewable Energy, Brussels, http://awsassets.panda.org/downloads/res_report_final.pdf

Subject index

Names of associations, campaigns, countries, decisions, directives, institutions, programmes, projects, scenarios are written in *italics*

Agency for the Cooperation of Energy Regulators (ACER) 90, 91, 92, 93, 97, 117
ALTENER 13, 14, 16, 17
Austria 4, 18, 34, 47, 48, 49, 75, 88, 94, 95, 106, 107, 140, 148
average annual growth rate (AAGR) 107, 118, 125, 149

barrier (for renewables deployment) 2, 10, 28, 36, 38, 39, 42, 47, 51, 53, 55, 63, 81, 82, 87, 92, 93, 106, 111, 113, 120, 137, 148, 156
    administrative barrier 10, 47, 156
    authorization procedure 83, 84
Belgium 18, 29, 34, 47, 48, 49, 88, 94, 95, 107
benefits (of sustainable/renewable energy) 3, 9, 12, 16, 27, 39, 42, 50, 51, 54, 55, 71, 94, 96, 97, 101, 103, 104, 105, 117, 118, 119, 121, 131, 135, 136, 137, 138, 147, 150, 151, 155
biofuels 10, 15, 18, 19, 20, 22, 23, 29, 31, 32, 36, 39, 40, 41, 42, 106, 108, 118, 123, 127, 128, 131, 138, 142, 158
    advanced biofuels 108, 141, 142
    blending (mandate) 23, 41, 131
    conventional biofuels 141, 142
    first generation biofuels 31, 40, 108
    second generation biofuels 23, 31, 40, 42, 108
Biofuels Directive 10, 19, 22, 29, 32, 36, 39, 158
biomass 4, 9, 10, 15, 16, 17, 18, 20, 22, 23, 25, 31, 40, 41, 42, 105, 106, 107, 108, 110, 113, 118, 126, 127, 128, 131, 137, 138, 141, 153, 159
    biogas 15, 18, 106, 130, 131, 159, 160
    bioliquids 41, 42, 142
Brazil 1, 3, 131, 142
bridging technology 28, 120
Bulgaria 34, 47, 48, 49, 94, 95, 107, 148

Campain for Take-off (CTO) 10, 16, 17
Carbon Capture and Storage (CCS) 10, 11, 24, 28, 67, 101, 112, 113, 114, 115, 116, 117, 122, 123, 124, 125, 135, 141, 144, 145, 153, 154, 155
    CCS-Directive 10, 24, 28, 29, 43, 144, 153
carbon price 25, 45, 59, 96, 106, 113, 114, 119, 122, 135, 142, 143, 144, 149, 154
    carbon intensity 14, 113
    carbon leakage 26, 60
    carbon-neutral 112
China 1, 3, 4, 131, 132, 145, 154
CHP (Combined Heat and Power) 37, 108, 120, 127

clean energy 3, 9, 11, 38
climate and energy framework (of the EU) 5, 42, 102, 135, 136, 145, 146, 150
climate and energy package (of the EU, 2009) 6, 10, 19, 21, 24, 25, 29, 42, 45, 81, 101, 104, 142, 144, 146, 149, 151, 158
climate change 1, 2, 3, 9, 10, 14, 21, 23, 26, 27, 28, 31, 58, 63, 65, 71, 81, 96, 101, 112, 121, 131, 133, 147
    global warming 1, 2, 112, 153
    greenhouse gas emission 1, 3, 9, 11, 14, 22, 24, 26, 27, 41, 45, 51, 57, 58, 59, 60, 61, 67, 96, 101, 103, 108, 114, 121, 122, 126, 135, 137, 142, 144, 145
    climate protection 9, 21, 22, 23, 26, 27, 39, 43, 53, 64, 81, 97, 101, 117, 135, 145, 159, 160
    decarbonisation 67, 114, 115, 116, 117, 121, 122, 123, 124, 125, 137, 138, 142, 144, 151, 158
    international agreement 26, 27, 58, 149, 153
    international climate process 1, 2
    Kyoto Protocol 2, 9, 10, 21, 26, 27, 57, 58, 77, 78
    unilateral 3, 22, 27
competitiveness 14, 23, 27, 112, 123, 135, 136, 137, 147, 148, 149, 150, 157, 159
compliance (with a directive, an obligation, a target, a treaty etc.) 11, 21, 25, 27, 29, 32, 33, 35, 36, 38, 40, 57, 58, 64, 66, 76, 78, 126, 139, 142, 143, 155
consumer (of energy) 16, 20, 38, 54, 55, 56, 64, 70, 77, 81, 82, 84, 85, 86, 88, 94, 96, 111, 119, 137, 138, 139, 146, 147
    energy consumption 15, 20, 23, 29, 32, 33, 45, 46, 48, 50, 52, 54, 74, 104, 106, 113, 118, 121, 124, 125, 126, 127, 128, 132, 143
    vulnerable customer 84, 85
costs (of energy, of energy transition) 3, 5, 6, 11, 16, 17, 38, 35, 39, 67, 71, 73, 88, 92, 96, 97, 101, 103, 104, 105, 106, 108, 109, 110, 111, 113, 114, 115, 116, 118, 119, 120, 122, 124, 125, 128, 130, 137, 138, 140, 144, 145, 146, 151, 153, 155, 156, 157, 158, 159, 160
    external costs/externalities 6, 74, 96, 114, 119. 140, 153, 158
Council of European Energy Regulators (CEER) 90
Covenant of Mayors 131
crisis (economic and financial) 11, 48, 53, 58, 59, 60, 93, 128, 132, 140, 149
curtailment (of power plants) 39, 159

175

Sustainable Energy Developments

Series Editor: Jochen Bundschuh

ISSN: 2164-0645

Publisher: CRC/Balkema, Taylor & Francis Group

1. Global Cooling – Strategies for Climate Protection
   Hans-Josef Fell
   2012
   ISBN: 978-0-415-62077-2 (Hbk)
   ISBN: 978-0-415-62853-2 (Pb)

2. Renewable Energy Applications for Freshwater Production
   Editors: Jochen Bundschuh & Jan Hoinkis
   2012
   ISBN: 978-0-415-62089-5 (Hbk)

3. Biomass as Energy Source: Resources, Systems and Applications
   Editor: Erik Dahlquist
   2013
   ISBN: 978-0-415-62087-1 (Hbk)

4. Technologies for Converting Biomass to useful Energy –
   Combustion, gasification, pyrolysis, torrefaction and fermentation
   Editor: Erik Dahlquist
   2013
   ISBN: 978-0-415-62088-8 (Hbk)

5. Green ICT & Energy – From smart to wise strategies
   Editors: Jaco H. Appelman, Martijn Warnier & Anwar Osseyran
   2013
   ISBN: 978-0-415-62096-3

6. Sustainable Energy Policies for Europe – Towards 100% Renewable Energy
   Rainer Hinrichs-Rahlwes
   2013
   ISBN: 978-0-415-62099-4 (Hbk)